Study Guide for

Introduction to General, Organic, and Biochemistry

NINTH EDITION

Bettelheim

Brown

Campbell

Farrell

Prepared by

William Scovell

Bowling Green State University

BROOKS/COLE
CENGAGE Learning™

Australia • Brazil • Japan • Korea • Mexico • Singapore • Spain • United Kingdom • United States

For product information and technology assistance, contact us at **Cengage Learning Customer & Sales Support, 1-800-354-9706**

For permission to use material from this text or product, submit all requests online at **www.cengage.com/permissions** Further permissions questions can be emailed to **permissionrequest@cengage.com**

ISBN-13: 978-0-495-39118-0
ISBN-10: 0-495-39118-2

Brooks/Cole
10 Davis Drive
Belmont, CA 94002-3098
USA

Cengage Learning is a leading provider of customized learning solutions with office locations around the globe, including Singapore, the United Kingdom, Australia, Mexico, Brazil, and Japan. Locate your local office at: **www.cengage.com/international**

Cengage Learning products are represented in Canada by Nelson Education, Ltd.

To learn more about Brooks/Cole, visit **www.cengage.com/brookscole**

Purchase any of our products at your local college store or at our preferred online store **www.ichapters.com**

Printed in the United States of America
1 2 3 4 5 6 7 12 11 10 09

TO THE STUDENT

I hear, and I forget
I see, and I remember
I do, and I understand

"Ancient Chinese Proverb"

The aim of this Study Guide is simply to aid the student in increasing understanding of the material presented in Introduction to General, Organic and Biochemistry by Bettelheim, Brown, Campbell and Farrell, and to make studying more efficient. The general philosophy or approach put forth here should not only be useful in this course, but in many courses in which the learning of specific new concepts and the development of problem solving abilities is important. The initial focus of attention should be to determine the immediate objectives or aims of each chapter. Therefore study objectives will begin every chapter in order to point out aspects worthy of special emphasis in your studies.

Once the major objectives have been read over carefully, one must ask, "How best can I study to develop a firm understanding and a command of the new vocabulary, concepts and problem solving techniques associated with the material?" The approach to this end will be to define four study practices.

1. Each chapter in the text includes a summary of the key concepts in that chapter. It is essential that the meaning of these terms, in **bold type**, be mastered. These and additional terms of importance from the text will be listed in the Study Guide, either individually, or in cases in which two or more terms are related, will be listed together. This latter grouping will serve to emphasize that the similarities and differences between these terms should be clearly understood. The terms are basic to the understanding of the new material and will provide the very foundation of your understanding. Review them together with the objectives before reading the chapter. After reading the text, reread the objectives and define each term. To deepen your understanding of the terms, include a specific example in each case where this is appropriate. Refer to the Glossary in the text to provide clear definitions to all the terms. **Continually review** the definitions before each study session or until the responses are clear and spontaneous.

2. In each chapter, a few concepts provide the basis for grasping many of the sometimes apparently diverse, but interrelated subjects presented. In some Study Guide chapters, a selected number of concepts will be elaborated on in order to "pull together" some of the ideas and to deepen understanding of the text material.

3. An integral part of chemistry involves the solving of both mathematical and conceptual problems. The general thought processes and interrelationships helpful in reaching these solutions will be briefly outlined. These problems will, in some cases, include the detailed solutions of some of the problems in the text.

4. At the point at which the student feels confident that the material has been mastered, a **SELF-TEST** and the **QUICK QUIZ** will provide a means for evaluating the progress toward

reaching the major objectives in the chapter. Answers to all the questions and problems will be included, with explanations accompanying some of the answers.

SOME OBSERVATIONS AND A SUGGESTED STUDY APPROACH

1.   Typically students who obtain above average grades will read the chapter a minimum of twice, if not three times. I point it out only because I find many freshman (and some upper classmen also) are surprised to learn that this effort is actually the **norm** and to be sure, not unusual.

2.   Reading the chapter prior to the material being discussed in class usually affords the student the unique advantage of being able to clarify questions on the material at the time the professor is introducing them to the class. Unfortunately, for many students, this "pre-reading" often becomes very difficult, if not impossible. However, try it. You may find it so valuable that it may become a part of your personal study strategies.

3.   Take good notes in class and as soon after class as possible, **recopy the notes into another notebook in a more organized fashion**. Do this at the same time that you refer to your textbook and annotate your class notes with extra definitions or comments from the text. I have found that this has been especially helpful to many students.

4.   Knowledge of a subject begins with **memorizing** the meaning of new terms, concepts and equations. However, **understanding** the material goes far beyond the memorization of these elements in a "vacuum." It requires that the interrelationships between all these elements be clearly resolved. This requires a continual review of the material and a major input of time and effort into solving problems and answering focused questions. Therefore, it is important that you commit yourself to working as many problems as possible. There is no substitute for this.

I would like to outline a study procedure that many students have found helpful. This, of course, will not be for everyone. Each student must develop routines which serve him or her best. Try this approach if you wish. If it helps, use it.

1.   **Review** the chapter objectives and important new terms in the Study Guide. This will alert you to the key ideas and concepts and help to underline their importance before your reading.

2.   **Read** the (text) chapter with these points in mind.

3.   **Record** good notes in class and as soon after class as possible, rewrite the notes. Clarify all questionable points in the lecture notes **AT THIS TIME**, with the aid of your text. For those concepts that are not clear, **ask the professor immediately** after the next lecture or bring up this point at his or her next office hour.

4.   **Write out** the definitions of all important new terms.

5. **Review** any worked problems in the chapter, with the aim of defining the general thought processes involved in solving the problem.

6. Now, the assigned questions and problems at the end of each chapter in the text should be worked and the material in the Study Guide should be studied. The guide will be most useful in cases in which some additional explanation or worked examples are helpful. In some cases, simply rephrasing of text material may bring out the meaning more clearly.

7. **Reread** the chapter, the rewritten class notes, chapter objectives, important terms and equations and the focused review section.

8. Return to doing questions and problems that were initially unclear or difficult and then do additional ones.

9. **Do the SELF-TEST** and the **QUICK QUIZ** in the study guide.

I hope that as you embark on this "chemistry" journey, you will expand your horizons and even enjoy the experience. Good luck!

William M. Scovell
Jan., 2009

TABLE OF CONTENTS

"It is only the first step
which takes the effort"
   - Madame Marie Vichy-Deffand

1

Matter, Energy, and Measurement

CHAPTER OBJECTIVES

After studying Chapter 1 and working the assigned exercises in the text and the study guide, you should be able to do the following.

1. Distinguish between a fact, a hypothesis and a theory. Indicate how they are associated with the scientific method and the role that experiments play in this process.

2. Convert large and small numbers to an equivalent form in exponential notation. Carry out the reverse procedure.

3. Perform addition, subtraction, multiplication and division calculations using exponential notation. Evaluate the answers to insure that there is the proper number of significant figures.

4. List the word prefixes used in the metric system and their corresponding values.

5. Define temperature and indicate temperature scales in common use. Write the equations that show the relationship between (i) ^oC and ^oF; (ii) ^oF and ^oC (iii) ^oC and K. Perform conversions from one temperature scale to another.

6. Using the factor-label method, perform conversions between the metric and English units, English to metric units, in addition to conversions within the metric system.

7. List the three states of matter and the characteristics that are unique to each state.

8. Distinguish between density and specific gravity.

9.   Define energy and list six different forms of energy pointed out in this chapter.

10. State the law of Conservation of Energy.

11. Distinguish between heat and specific heat and indicate the units of each.

12. Using the rules for determining the number of significant figures, evaluate the number of significant figures in numbers listed throughout the chapters.

13. Define the important terms and comparisons in this chapter and give specific examples where appropriate.

IMPORTANT TERMS AND COMPARISONS

Chemistry	Matter and Energy
Chemical and Physical Properties	Chemical and Physical Changes
Scientific Method	Fact, Hypothesis, Experiment and Theory
Exponential Notation	Coefficient and Exponent
Significant Figures	Kilo- and Milli-
English vs. Metric System of Units	Celsius, Fahrenheit and Kelvin Scales
Mega- and Micro-	Centi-
Measured and Defined Numbers	Mass and Weight
Units of Length, Volume, Mass and Time	States of Matter
Density and Specific Gravity	Potential and Kinetic Energy
Law of Conservation of Energy	Temperature and Heat
Heat and Specific Heat	Calorie and Kilocalorie
Hypothermia and Hyperthermia	

FOCUSED REVIEW OF CONCEPTS

A.   Underline{Interrelationships}

  (i) $°F = 9/5(°C) + 32$

  (ii) $°C = 5/9(°F-32)$

  (iii) $K = °C + 273$

  (iv)   Density = mass/volume; $D = M/V$
          Density is a **physical quantity** and D has **units** (e. g., g/mL)

2

(v)   Specific Gravity (S.G.) =

<u>Density of substance of interest</u>
Density of water

S.G. = D_A/D_{H_2O}

S.G. is a **ratio** of densities and is a **unitless number**

(vi)   Heat (cal) = Specific heat (cal/g-deg) x mass (g) x (T_2-T_1)

It should be noted that both equations (i) and (ii) define the relationship between the temperature in degrees Celsius and in degrees Fahrenheit. Therefore, you should be able to rearrange equation (i) to arrive at equation (ii). Let's do it.

Write eqn (i)

$$°F = 9/5 \ (°C) + 32$$

Subtract 32 from each side

$$°F-32 = 9/5(°C)+32-32$$

$$(°F-32) = 9/5(°C)$$

Multiple each side by 5/9

$$5/9(°F-32) = (5/9)(9/5)(°C)$$

This is eqn (ii)

$$5/9(°F-32) = °C$$

Equation (iii) should be put to memory because many calculations to be performed in subsequent chapters (e.g., Chapter 6) will require that the temperature be expressed in degrees Kelvin (K).

Density is an intensive property of a substance and has units of mass/vol, which although usually expressed in g/mL, can be expressed in others units (i.e., lb/in.^{3}, etc.). Recall also that **in**tensive properties are those which are **in**trinsic characteristics of the substance (i.e., density) and therefore do not depend on the quantity of the substance.

Specific gravity is a ratio of the density of any substance to the density of water at 20°C. Therefore, specific gravity is simply a number, with no units.

B. Significant Figures

A scientist must evaluate the number of significant figures in every **measurement** and in every **calculation** using measured numbers. The number of significant figures is defined as the number of digits that are known with certainty.

The rules are the following. Underlined numbers are not significant.

1. All digits written down are significant, except in the case of zeros (19578).
    a. Zeros that come before the first non-zero digit are not significant (0.456 or 0.089)
    b. Zeros that come after the last non-zero digit are significant if the number is a decimal. However, if the number is a whole number (no decimal point), they may or may not be significant (3.200; however, the zeros in 32,000 may or may not be significant)
2. In multiplying or dividing numbers, the **final answer** should have the same number of significant numbers as there are in the number with the fewest significant numbers.
3. In adding or subtracting a group of numbers and arriving at the number of significant figures in the answer, the number of significant numbers in the individual numbers does not matter. The answer must contain the same number of decimal places as the number in the listing with the fewest decimal places.
4. In rounding off numbers, if the first digit dropped is a 5, 6, 7, 8 or 9, raise the last digit kept to the next higher number. In all other cases, do not raise the last digit.
5. Counted or defined numbers, such as 4, the number of sides on a square, are treated as if they have an infinite number of zeros following the decimal point.

PROBLEM SOLVING METHODS

A. Exponential Notation

It is important to be able readily to express large and small numbers in exponential or scientific notation. This requires that the number be written as a **coefficient,** with a value between 1 and 10, **multiplied** by 10 raised to some power. For example,

$100 = 1 \times 10^2 = 1 \times (100)$ The solid line underlines the number of places the decimal point has been moved in each case.

$20{,}000 = 2 \times 10^4 = 2 \times (10{,}000)$

$5{,}200{,}000 = 5.2 \times 10^6 = 5.2 \times (1{,}000{,}000)$

Note in these examples that when the LARGE number (greater than 1) is changed to exponential notation form, the decimal point is moved to the LEFT by 2, 4 or 6 places, respectively. The **positive** exponent simply expresses the number of places that the decimal has been moved to the left.

Similarly, in changing these SMALL numbers (less than one) to exponential notation, the decimal point is moved to the RIGHT by 3, 5 or 2, respectively. The **negative** exponent indicates that the number is smaller than 1.

4

$0.\underline{001} = 1 \times 10^{-3} = 1 \times (0.\underline{001})$

$0.\underline{000043} = 4.3 \times 10^{-5} = 4.3 \times (0.\underline{00001})$

$0.\underline{0200100} = 2.00100 \times 10^{-2} = 2.001 \times (0.\underline{01})$

a. MOVING THE DECIMAL POINT in Numbers Already in Exponential Notation.

One of the most common difficulties that students encounter occurs when doing multiple step calculations. Midway through the calculation, numbers such as the following may appear.

$16,743 \times 10^{14}$ 0.00713×10^6 0.0000431×10^{-6}

To change these numbers to proper form (i.e., the coefficient must be between 1 and 10), the decimal point must be moved. Which way is always the question!

The rule to remember is:

IF THE **COEFFICIENT IS TO BE DECREASED** in magnitude, the **EXPONENT MUST BE INCREASED ACCORDINGLY.**

IF THE **COEFFICIENT IS TO BE INCREASED**, the **EXPONENT IS DECREASED.**

For example:

$16,\underline{743} \times 10^{14}$ $\rightarrow$ 1.6743×10^{18}
> **coefficient decreased**, decimal moved **4** places
> **exponent increased by 4**

$0.\underline{007}13 \times 10^6$ $\rightarrow$ 7.13×10^3
> **coefficient increased**, decimal moved **3** places
> **exponent decreased by 3**

$0.\underline{0000}431 \times 10^{-6}$ $\rightarrow$ 4.31×10^{-11}
> **coefficient increased**, decimal moved **5** places
> **exponent decreased by 5**

Exercise I: Carry out these manipulations to develop skill in these conversions.

Number   Exponential Notation (with coefficient between 1 and 10)

1. 0.00102×10^1 _____

2. 632×10^{-14} _____

3. 0.00700×10^{43} _____

4. $13,642,000 \times 10^{-1}$ _____

5. 11×10^{23} _____

6. $0.0000006 \times 10^{-14}$ _____

7. 137×10^{-10} _____

8. 0.006×10^{0} _____

9. $7,423,000,000 \times 10^{17}$ _____

10. 0.4361×10^{31} _____

In the frequently used manipulations with numbers in exponential notation outlined below, a TWO-STEP procedure can be used in each case.

b.   ADDITION OR SUBTRACTION in Exponential Notation

Procedure: (i)   change the numbers so all have the same exponent,
                       irrespect of the size of the coefficient
              (ii) Add (or subtract) the coefficients; the exponent remains the same

Ex.   a. 230×10^{-8} (i) 23×10^{-7}
           $\underline{6.10 \times 10^{-5}} \rightarrow$ $\underline{610 \times 10^{-7}}$
                                    (ii) $\downarrow$ $\downarrow$
                                                $633 \times 10^{-7} = 6.33 \times 10^{-5}$

      b. 340×10^{6} (ii) 340×10^{6}
        - $\underline{220 \times 10^{5}} \rightarrow$ $-\ \underline{22 \times 10^{6}}$
                                    (ii) $\downarrow$ $\downarrow$
                                                $318 \times 10^{6} = 3.2 \times 10^{8}$

c.   MULTIPLICATION in Exponential Notation

Procedure: (i)   Multiple the coefficients directly
              (ii) Algebraically **add** the exponents
              (iii) Combine coefficients and exponent
              (iv) If necessary, change the coefficient to a value between 1 and 10
                     and make the corresponding change in the value of the exponent.

6

Ex. a. $(6.0 \times 10^{-14})(3.0 \times 10^6)$ (i) $\rightarrow$ 18

(ii) $\rightarrow$ x 10^{-8}

(iii) 18 x 10^{-8}

(iv) <u>1.8 x 10^{-7}</u>

b. $(1 \times 10^{-14})(7 \times 10^{-2})$ (i) $\rightarrow$ 7

(ii) $\rightarrow$ x 10^{-16}

(iii & iv) <u>7 x 10^{-16}</u>

d. DIVISION in Exponential Notation

Procedure: (i) Divide the coefficients directly

(ii) Algebraically subtract the exponents

(iii) Combine

(iv) If necessary, change coefficient to a value between 1 and 10 and make the corresponding change in the value of the exponent.

Ex. a. $\dfrac{3 \times 10^{-7}}{1 \times 10^7}$ (i) $\rightarrow$ 3

(ii) $\rightarrow$ x 10^{-7-7}

(iii & iv) <u>3 x 10^{-14}</u>

b. $\dfrac{3 \times 10^7}{1 \times 10^{-7}}$ (i) $\rightarrow$ 3

(ii) $\rightarrow$ x $10^{7-(-7)}$

(iii & iv) <u>3 x 10^{14}</u>

c. $\dfrac{2.14 \times 10^{-3}}{8.60 \times 10^{+4}}$ (i) $\rightarrow$ 0.249

(ii) $\rightarrow$ x 10^{-3-4}

(iii & iv) 0.249 x 10^{-7} = <u>2.49 x 10^{-8}</u>

d. $\dfrac{5.91 \times 10^{-4}}{1.61 \times 10^{-12}}$ (i) $\rightarrow$ 3.67

(ii) $\rightarrow$ x 10^{-4+12}

(iii & iv) <u>3.67 x 10^8</u>

B. Unit Conversions

(i) All conversions within the metric system differ by some power of 10 as reflected in the prefixes. For example, centi = 10^{-2}; milli = 10^{-3}; micro = 10^{-6}; nano = 10^{-9}. These conversions are accomplished simply by moving the decimal point. Memorize these prefixes and those in Table 1.2 (text).

The strategy in these problems is to multiply the starting physical quantity (a number with a unit) by a conversion factor that will both change the number and also its units.

7

AT EACH STEP IN THE CONVERSION, ONE UNIT MUST CANCEL OUT. AT NO POINT IN THE PROCESS ARE THERE MORE UNITS THAN YOU STARTED WITH. This conversion process is continued until the desired unit (or units) is (are) obtained. In each case, clearly understand what the initial units are and also what the units are you want to convert to. Use the conversion factors listed in Table 1.3 (text)

Ex. 1   Change 2 ft to inches.   Start with **feet;**
                                           Convert to **inches**

$$\left(\frac{2 \text{ ft}}{1}\right)\left(\frac{12 \text{ in.}}{1 \text{ ft}}\right) \qquad \rightarrow \qquad \frac{24 \text{ in.}}{1} \quad = \quad 24 \text{ inches}$$

starting   conversion
physical   factor
quantity

Note that in this one-step conversion, 2 units cancel (ft) and only one unit (in., the desired unit) remains. If the correct conversion factor had been used, but incorrectly inverted, **no units would have canceled**. For example:

$$\left(\frac{2 \text{ ft}}{1}\right)\left(\frac{1 \text{ ft}}{12 \text{ in.}}\right) \qquad \rightarrow \qquad \frac{2 \text{ ft}^2}{12 \text{ in}}$$

Units do not cancel. This **incorrect** set-up produced more units (ft^2/in) than in the starting quantity. This is your signal that the conversion factor is inverted.

Ex. 2   Convert 3.2 milliliters to nanoliters.

Start with **milliliters;** convert to **nanoliters**. Depending on the conversion factors you know, this problem and many others, can be done in one or perhaps a few steps.

Route 1: $\dfrac{3.2 \text{ mL}}{1} \quad \dfrac{10^6 \text{ nL}}{1 \text{ mL}} \quad = \quad 3.2 \times 10^6 \text{ nL}$

Route 2: $\dfrac{3.2 \text{ mL}}{1} \quad \dfrac{10^3 \text{ } \mu\text{L}}{1 \text{ mL}} \quad \dfrac{10^3 \text{ nL}}{\mu\text{L}} \quad = \quad 3.2 \times 10^6 \text{ nL}$

Although $1 \text{ nL} = 10^{-9} \text{ L} = 10^{-6} \text{ mL} = 10^{-3} \text{ } \mu\text{L}$, you should note that the conversion factors are:

$\dfrac{10^9 \text{ nL}}{\text{L}}, \quad \dfrac{10^6 \text{ nL}}{\text{mL}}, \quad \dfrac{10^3 \text{ nL}}{\mu\text{L}}.$

In each case, the exponents are **positive** to show the number of nanoliters contained in the larger volume unit.

8

Exercise II: Convert the quantities shown below into a metric equivalent. Since all the quantities are in the metric system, the conversion requires only moving the decimal to the left or the right.

1.   1.32 L   _____ mL   _____ μL   _____ nL

2.   6.32 mg   _____ g   _____ kg   _____ Mg

3.   0.039 cm   _____ m   _____ km   _____ μm

4.   0.36 μs   _____ ms   _____ s   _____ ns

     (ii)   General conversions that are not within the metric system require the same procedure. The conversion factors used, however, do not simply move the decimal point (i.e., the exponent), but change both the coefficient and the exponent in the starting number.

Ex. 1   How many nickels are in $363.00?
           Start with **dollars**; convert to **nickels**.

$$\frac{363 \text{ dollars}}{1} \quad \frac{20 \text{ nickels}}{1 \text{ dollar}} \quad = \quad (363)\,(20) \text{ nickels} = 7{,}260 \text{ nickels}$$

Ex. 2   The density of water is 1 g/cc. What is this equivalent to in terms of lb/in.^{3}?

Solution:

This is a multistep conversion in which 2 units are to be converted.

$$g \quad \rightarrow \quad lb.$$

$$cc \quad \rightarrow \quad in.^3$$

These problems are done by **completely converting one unit**, and then the other.

Recall 1 mL = 1 cc = 1 cm^3

$$\left(\frac{1.0 \text{ g}}{\text{cm}^3}\right)\left(\frac{1.0 \text{ lb}}{454 \text{ g}}\right)\left(\frac{2.54 \text{ cm}}{1 \text{ in.}}\right)\left(\frac{2.54 \text{ cm}}{1 \text{ in.}}\right)\left(\frac{2.54 \text{ cm}}{1 \text{ in.}}\right) = \frac{(2.54)^3 \text{ lb}}{454 \text{ in.}^3}$$

conversion   conversion of
from g to lb   cm^3 → in^3

$$= 3.6 \times 10^{-2}\left(\frac{\text{lb}}{\text{in.}^3}\right)$$

Ex. 3   Which of the following is the largest mass, and which is the smallest?

a) 3.5×10^{-8} kg   c) 4×10^3 mg

b)   27 g   d) 3.4×10^6 μg

Solution:

The main difficulty in determining which quantity is the largest and which is the smallest occurs because each mass is expressed in a different unit. The first task then is to convert all the masses to the same unit. It is **not important which units are finally compared** (kg with kg, g with g or other units). What is important is that **the numbers all contain the same units**. The question here will be answered by comparing all the masses in units of **μgs**.

a. $\left(\dfrac{3.5 \times 10^{-8} \text{ kg}}{1}\right)\left(\dfrac{10^3 \text{ g}}{1 \text{ kg}}\right)\left(\dfrac{10^6 \text{ μg}}{1 \text{ g}}\right)$ $=$ $3.5 \times 10^{-8+3+6}$ μg

$=$ 3.5×10^1 μg

b. $\left(\dfrac{27 \text{ g}}{1}\right)\left(\dfrac{10^6 \text{ μg}}{1 \text{ g}}\right)$ $=$ 2.7×10^7 μg

c. $\left(\dfrac{4 \times 10^3 \text{ mg}}{1}\right)\left(\dfrac{10^3 \text{ μg}}{1 \text{ mg}}\right)$ $=$ 4×10^6 μg

d. 3.4×10^6 μg $=$ 3.4×10^6 μg

The comparison is now between all masses expressed in μg.

a. 3.5×10^1 μg

b. 2.7×10^7 μg or 27×10^6 μg

c. 4×10^6 μg

d. 3.4×10^6 μg

$3.5 \times 10^{+1}$ μg $<$ 3.4×10^6 μg $<$ 4×10^6 μg $<$ 2.7×10^7 μg

————————————— increasing mass —————————————>

Therefore,
        a. has the smallest mass (35 μg)

        b. has the largest mass (2.7×10^7 μg or 27,000,000 μg)

10

Work this problem again for practice and convert all the masses to a different common unit.

Exercise III: Perform the following conversions from the English to the metric system of units or vice versa. These conversions require the use of conversion factors and the factor label method. Refer to Table 1.3 (text) for the other widely used conversion factors.

1.   1.0 ft   _____ in   _____ cm   _____ mile

2.   300. g   _____ oz   _____ lb   _____ μg

3.   6.3 fluid oz. _____ μL   _____ L   _____ qt

Ex.4   If the density of air is 1.25×10^{-3} g/cc, what is the mass of the air in a room that is 5.0 meters long, 4.0 meters wide and 2.2 meters high?

Solution:

One approach to this problem is the following:

(i)   The density is known in g/cc.

(ii)   Determine the volume of the room in meters cubed (m^3) and then convert this volume to cc.

(iii)   Knowing the density now in g/cc and the volume in cc, calculate the mass of the air in the room in grams.

(i) $D = 1.25 \times 10^{-3}$ g/cc

(ii)   Determine the volume

        Volume = length x width x depth = (5.0 m) (4.0 m) (2.2 m) = 44 m^3

        Conversion of 44 m^3 to cc

$$\left(\frac{44\ m^3}{1}\right)\left(\frac{10^2\ cm}{1\ m}\right)\left(\frac{10^2\ cm}{1\ m}\right)\left(\frac{10^2\ cm}{1\ m}\right) = 44 \times 10^6\ cm^3\ or\ 4.4 \times 10^7\ cc$$

(iii)   Determine the mass of air.

        $D = M/V$ $M = (D)(V)$
                         $= (1.25 \times 10^{-3}\ g/cc)\ (4.4 \times 10^7\ cc)$

        $M = 5.5 \times 10^4$ g

11

THE CONCEPT OF SPECIFIC HEAT

The specific heat of a substance is an intensive property of the substance - that is, it does not depend on the quantity of the substance. It is a measure of how much energy (heat) is required to increase the temperature of one gram of a substance by 1°C.

Ex. 1   Consider the specific heat of 3 substances.

specific heat (cal/g-deg)

water	1.00
iron	0.11
lead	0.038

a.   Which substance requires the most heat to increase 1 g of the substance by 1°C?

b.   If equivalent amounts of heat (for ex., 100 cal) are added to 1 g of each substance, which substance will experience the largest increase in temperature? Which will experience the least increase?

c.   If 60 cals. are added to 6 g of iron at 20°C, what will the final temperature be?

Solution:

a.   Using the definition of specific heat directly, the values indicate that the specific heat for water requires 1.00 cal to increase 1 g of water by 1°C. The specific heat value for iron is about one tenth the value for water, while the value for lead is smaller yet. Therefore, by definition and simple comparison, water requires the most heat to raise its temperature by 1°C.

b.   An equivalent amount of heat added to 1 g of each substance would increase the temperature of lead the most, while the increase in water would be the least. This is exactly the opposite order as observed in part a.

c.   **Heat** absorbed by a substance will produce an increase in **temperature**. The magnitude of the increase will depend on the **specific heat** (S.H.) of the substance and the **mass**. As shown in the section 1.9 in the text, the equation relating these quantities is:

$$\text{heat} = (\text{S.H.}) \times (m) \times (T_2 - T_1)$$

This equation has 5 quantities (heat, S.H., m, T_2, T_1). In this problem and in others, you must first determine (i) which quantities are given and (ii) which quantity is to be determined (i.e., the unknown). Because only one quantity in this problem is to be determined, 4 quantities must be known.

Known quantities: $\qquad$ heat $=$ 60 cal

$$m = 6\,g$$

$$\text{S.H. (iron)} = 0.11\ cal/g\,^0C$$

$$T_1 = 20°C$$

$$T_2 \text{ is to be determined}$$

Inserting these quantities into the equation

$$60\ cal = (0.11\ cal/g\,^0C)\,(6g)\,(T_2\text{-}20°)$$

Rearrange this equation by dividing both sides by (0.11 cal/g-deg) (6 g) and then canceling on the right side to yield

$$\frac{60\ cal}{(0.11\ cal/g\,^0C)\,(6\ g)} = (T_2\text{-}20°C)$$

$$91° = T_2\text{-}20$$

Finally, add 20 to both sides to arrive at the final temperature, T_2.

$$111°C = T_2$$

Ex. 2   If 425 cal of heat are added to a sample of 52 g of copper at 25°C, calculate the final temperature?

$$\text{heat} = (\text{S.H.}) \times (m) \times (T_2\text{-}T_1)$$

Given:

$$\text{heat} = 425\ cal$$

$$m = 52\ g$$

$$\text{S.H. (copper)} = 0.092\ \text{(Table 1.3, text)}$$

$$T_1 = 25°C$$

$$T_2 \text{ is to be determined}$$

$$425\ cal = 0.092\ (cal/g\,^0C) \times (52\ g) \times (T_2\text{-}25°C)$$

13

$$\frac{425 \text{ cal}}{(0.092 \text{ cal/g }^{\circ}\text{C}) (52 \text{ g})} = T_2\text{-}25^{\circ}\text{C}$$

$$89^{\circ} = T_2\text{-}25^{\circ}\text{C}$$

$$114^{\circ} = T_2$$

SELF-TEST QUESTIONS

<u>MULTIPLE CHOICE</u>. In the following questions, select the correct answer from the choices listed. In some cases two or more answers will be correct.

1. Indicate which of the following are extensive properties of a substance - that is, a property that depends on the quantity of the substance.

   a.   mass
   b.   density
   c.   length

   d.   specific heat
   e.   kinetic energy
   f.   melting point

2. Indicate whether each statement is a fact, hypothesis or a theory.

   a.   Smoking cigarettes is detrimental to your health.
   b.   The ocean water at Miami, Florida contains more salt than the water in Galvaston, Texas.
   c.   A micrometer (um) is larger than a nanometer (nm).
   d.   Judy received a higher grade on the exam because she studied longer.

3. Which of the following are forms of matter?

   a.   thought
   b.   nucleus of an atom
   c.   a vacuum in space

   d.   a line drawn on a paper
   e.   a shadow

4. The number 2.2×10^{-4} is equivalent to:

   a.   0.0022
   b. 22×10^{-5}

   c. 22×10^{-3}
   d. 0.022×10^{-2}

5. The number 1×10^{-4} is equivalent to:

   a. $\dfrac{1}{1 \times 10^4}$
   b. 10×10^{-3}

   c.   0.0001
   d. 100×10^{-6}

6. Which of the following numbers can be added directly in exponential notation?

    a. 3.26×10^{-4} c. 32.6×10^{-3}
    b. 600×10^{-2} d. 0.40×10^{-4}

7. Multiplication of $(3 \times 10^{-7}) (1 \times 10^6) (2 \times 10^0)$ equals:

    a. 3×10^{-1} c. 6
    b. 6×10^{-1} d. 60×10^{-3}

8. The answer to the problem $\dfrac{93 \times 10^4}{3.0 \times 10^{-4}}$ is:

    a. $31 \times 10^{+8}$ c. 13
    b. 31 d. 31×10^0

9. The answer to the problem $\dfrac{(3 \times 10^{-7})(5 \times 10^2)}{[3 \times 10^1 + 20]}$ is:

    a. 5×10^{-4} c. 3×10^{-4}
    b. 3×10^{-6} d. 0.005

10. Which of the following are equivalent to 50 mL?

    a. 0.5L   c. 0.05 L
    b. 50 cc   d. 5×10^4 μL

11. Which of the following are equivalent to 0.5 ms?

    a. 500 μs   c. 8×10^{-6} min
    b. 5×10^{-4}s   d. 5000 μs

12. A degree on the Celsius scale is:

    a. larger than the degree on the Fahrenheit scale
    b. is the same size as the degree on the Kelvin scale
    c. 9/5 larger than the Fahrenheit scale
    d. none of the above

13. The temperature of -40°C corresponds to:

    a. 313 K   c. -40°F
    b. 233 K   d. -73°F

15

14. The temperature of 0 °K corresponds to:

    a.   273°C   c.   -459.4°F
    b.   absolute zero   d.   32°F

15. A density of 1.5 g/cc is equivalent to:

    a.   1500 g/L   c. 3.3×10^{-3} lb/mL
    b.   13.4 lb/ft^3 d.   6

16. How much energy is required to warm 15.0 g of ethanol from 4°C to 25°C? The specific heat
    of ethanol is 0.590 cal/g °C.

    a.   158 cal   c.   0.185 kcal
    b.   185 cal   d.   37 cal

17. Which of the following ratios may be used directly as a conversion factor in converting
    microliters (µL) to liters (L)?

    a. $\dfrac{1 \text{ gal}}{3.785 \text{ L}}$ b. $\dfrac{1 \text{ L}}{10^{+6} \text{ uL}}$ c. $\dfrac{10^{6} \mu \text{L}}{1 \text{ L}}$ d. $\dfrac{10^{-6} \text{ uL}}{1 \text{ L}}$

18. Indicate the number of significant figures in the following measured numbers.

    a.   0.02   b.   32.00032   c. 0.00011   d.   120,000

COMPLETION. Write the word, phrase or number in the blank that will complete the statement
or answer the question.

1. Matter is defined as _____.

2. Scientists proposed new _____ to be tested and continually examine the validity of
   _____ in more depth.

3. In the number, 4×10^{-4}, the number 4 is called the _____, while -4 is the
   _____.

4. Another way of writing the following numbers is:

    a. $\dfrac{1}{10^{-4}}$ = _____   c. 43×10^{6} = _____
    b.   0.001 = _____   d.   9,430,000,000 = _____

16

5. The number $87{,}000 \times 10^{-16}$ is the same as 8.7×10^{-n} in which n equals
   _____.

6. SI is the abbreviation for _____. This system is based on the
   _____.

7. In the SI, the base unit of length is the _____, which is abbreviated as
   _____; the base unit of volume is _____, which is abbreviated as
   _____.

8. A kilometer is _____ times longer than a centimeter.

9. Scientists use a balance to experimentally determine the _____ of a substance.

10. The base unit for _____ is the same in the English and metric systems. This
    abbreviation is _____.

11. The size of the degree on the Kelvin scale is _____ to that on the Celsius scale.

12. A temperature of 300°C corresponds to _____ K.

13. A temperature of 37°C is equal to _____ °F.

14. Absolute zero is defined as _____ K. At this temperature, the motion of molecules
    _____.

15. There are _____ L in 6.7 gallons of wine.

16. The three states of matter are _____, _____ and _____.

17. To change a liquid to a gas, _____ must be added to the liquid.

18. To change a liquid to a solid, energy must be _____ from the liquid.

19. For the three states of matter, the order of compressibility is
    _____ > _____ > _____.

20. Oil spilled by large tankers floats on the ocean because oil is _____ than ocean water.

21. The density and the specific gravity of a substance differ in that _____.

22. Density and specific gravity are _____ properties of a substance.

23. The density of a substance _____ as the temperature is increased.

24. A hydrometer is used to measure the _____ of a liquid.

25. Energy is defined as _____.

17

26. Kinetic energy is the energy of _____, while potential energy is _____ energy.

27. The law of conservation of energy states that _____.

28. A high value for the specific heat of a substance indicates that _____.

29. Substance 1 has a specific heat of 3.0 cal/g ^{0}C, while substance 2 has a specific heat of 1.0 cal/g ^{0}C at 20°C. The addition of an equivalent amount of heat to 10 grams of either substance will result in substance _____ increasing more in temperature.

30. In changing a large number into exponential notation form, the decimal point is moved to the _____ by n places and the exponent is equal to _____.

31. A temperature of 150 K is equal to _____ on the Fahrenheit scale.

32. A volume of 131 mL of water weighs _____.

33. A 15.0 mL volume of liquid with a specific gravity of 1.75 has a mass of _____.

34. A volume of a clear blue liquid with a density of 0.89 g/cc weighs 43 g. The volume is _____. Its specific gravity is equal to _____.

35. A copper atom weighs 2.3×10^{-25} lb. Its weight in milligrams is _____, and its corresponding weight in kilograms is _____.

36. A drop of rain has a volume of about 50 microliters. This is equivalent to _____ quarts (1 quart = 0.946 L).

37. A single bond between two carbon atoms has a length of 1.54 Angstroms (1 Angstrom = 1×10^{-10} m). This bond length is equivalent to _____ in.

38. A substance is found to have a density of 2.40 g/cc and weighs 100. g. Its volume is _____.

39. At 20°C, the specific gravity of a solid substance A is 2.46, while the density of a liquid B is 2.0 g/cc. If solid A is placed in liquid B, it will _____.

40. A block of Blasa wood has dimensions of 2.00 cm x 6.00 cm x 1.50 m and a density of 0.120 g/cc. Its mass is _____.

41. Express the following quantities using the metric (SI) units indicated.

    a.   34 cm = _____ m

    b. 1×10^3 g = _____ µg

18

c. 2.3×10^3 g/cc = _____ mg/L

d.   6.3 cal/g ^{0}C = _____ kcal/mg ^{0}C

42.   While solids have a definite _____ and _____, a liquid has only a definite _____., while a _____ has neither.

43.   The _____ of an object is independent of its location on earth, while the _____ of an object is dependent on its location.

44.   The _____ of a substance is measured on a laboratory balance.

45.   The density of a substance usually _____ with increasing temperature because the _____ increases with temperature, while the _____ does not change with temperature.

46.   A billion seconds is equivalent to ca. _____ years. A billion seconds ago, most students in this class were probably _____ born yet.

47.   Indicate the number of significant figures in the following

    a.   2.00   _____

    b.   11,000   _____

    c.   0.0043   _____

    d. 5.32×10^4 _____

    e. 5.0×10^{-3} _____

48.   Carry out the following operations and report the correct number of significant figures in the answer.

    a.   14.00   b.   3,640.64
           1.042   -   13.1
          307.

    c. 0.37×41.56 =   _____

    d. $\dfrac{18.4}{3.468}$ =   _____

    e. $\dfrac{0.1340 \times 6.20}{146.00}$ =   _____

49. When heat is added to a substance, the temperature of the substance increases, with the magnitude of the temperature rise depending on the specific heat of the substance. Distinguish between the terms temperature, heat and specific heat by writing out the definition for each.

<u>QUICK QUIZ</u>

Indicate whether the statements are true or false. The number of the question refers to the section of the book that the question was taken from.

1.4 How Do We Make Measurements?
(a) All the units in the metric system differ from those in the English system.
(b) The SI units are exactly the same as those in the metric system.
(c) Units of length in the metric system differ from each other by powers of 10.
(d) Volume measurements are done in metric units in chemistry and in medicine.
(e) Mass and weight are two different names for the same thing.
(f) Body mass is an important consideration in determining drug dosage.
(g) In the Celsius temperature scale, the freezing point of water is at 0°C.
(h) The size of the Kelvin degree is the same as the size of the Fahrenheit degree.
(i) Absolute zero in temperature is 0 K (zero degrees Kelvin).

1.5 What Is A Handy Way to Convert from One Unit to Another?
(a) It is not particularly important to keep track of units in chemical calculations.
(b) When we multiply numbers, we also multiply units.
(c) When we divide numbers, we also divide units.
(d) It is safe to accept the answer shown on a calculator without checking further.
(e) If the units of the answer to a calculation are not the units we are looking for, the calculation must be wrong.
(f) If the units of the answer to a calculation are the units we are looking for, the calculation must be right.

1.8 How Do We Describe the Various Forms of Energy?
(a) Energy is the capacity to do work.
(b) Kinetic energy is stored energy.
(c) Potential energy is energy of motion.
(d) Kinetic energy can be converted to potential energy, but not vice versa.
(e) A bullet has the same kinetic energy at rest or when shot out of a gun
(f) Two objects moving at the same speed always have the same kinetic energy

1.9 How Do We Describe Heat and the Ways in Which It Is Transferred?
(a) Heat and temperature are the same.
(b) The calorie is a unit of heat.
(c) The joule is not a unit of heat.
(d) All substances require the same amount of heat to raise the temperature of one gram of the substance 1°C.
(e) If the sun shines on a swimming pool and on an iron bar for 5 hours, both will end up being the same temperature.

20

ANSWERS FOR FOCUSED REVIEW EXERCISES

Exercise I

1. 1.02×10^{-2}
2. 6.32×10^{-12}
3. 7.00×10^{40}
4. 1.3642×10^6
5. 1.1×10^{24}
6. 6×10^{-21}
7. 1.37×10^{-8}
8. 6×10^{-3}
9. 7.423×10^{26}
10. 4.361×10^{30}

Exercise II

1. 1.32×10^3 mL; 1.32×10^6 μL; 1.32×10^9 nL
2. 6.32×10^{-3} g; 6.32×10^{-6} kg; 6.32×10^{-9} Mg
3. 3.9×10^{-4} m; 3.9×10^{-7} km; 3.9×10^2 μm
4. 3.6×10^{-4} ms; 3.6×10^{-7} s; 3.6×10^2 ns

Exercise III

1. 12 in; 30 cm; 1.9×10^{-4} miles
2. 10.5 oz; 0.661 lb; 3.00×10^8 μg
3. 1.9×10^5 μL; 1.9×10^{-1} L; 2.0×10^{-1} qts

ANSWERS TO SELF-TEST QUESTIONS

MULTIPLE CHOICE

1. a, c, e
2. a (fact); b & d(hypothesis); c(fact)
3. b, d
4. b, d
5. a, c, d
6. a, d
7. b
8. a
9. b
10. b, c, d
11. a, b, c
12. a, b, c
13. b, c
14. b, c
15. a, c
16. b, c
17. b; in this conversion, L must be in the numerator
18. a. one   b. seven   c. two
    d. ambiguous. Can be two or as many as six.

COMPLETION

1. anything that has mass and also occupies space
2. hypotheses, theories
3. coefficient, exponent

4. a. 1×10^4

    b. 1×10^{-3}

    c. 4.3×10^7

    d. 9.43×10^9
5. 12
6. System International, Metric System
7. meter, m, liter, L
8. 10^5
9. mass
10. time, s
11. identical
12. 573 K
13. 98.6°F
14. 0 K or -273°C, ceases
15. 25 liters
16. solid, liquid, gas
17. heat energy
18. taken away
19. gas > liquid > solid
20. less dense
21. density is a physical quantity and has units, while specific gravity is a ratio of densities and is unitless
22. **in**tensive
23. decreases
24. specific gravity
25. the capacity to do work
26. motion; stored
27. energy can neither be created or destroyed
28. the input of large amounts of energy or heat is necessary to increase the temperature of the substance
29. substance 2
30. left, plus n
31. -189°F
32. 131 g
33. 26.2 g
34. 48 cc (mL), 0.89
35. 1.0×10^{-19} mg, 1.0×10^{-25} kg
36. 5.3×10^{-5} qt
37. 6.06×10^{-9} in
38. 42 cc (mL)
39. sink
40. 216 g
41. a. 0.34 m

    b. $1 \times 10^{+9}$ μg

    c. 2.3×10^9 mg/L

    d. 6.3×10^{-6} cal/mg-deg

22

42. shape, volume, volume, gas
43. mass, weight
44. mass
45. decreases, volume, mass
46. 31.7 years. In 2005, a billion seconds ago would be about the year 1973. Most students in this class were not born yet.
47. a. 3
    b. at least 2; if all figures are significant, then 5
    c. 2
    d. 3
    e. 2
48. a. 322
    b. 3627.5

    c. 15
    d. 5.31
    e. 5.69×10^{-3}
49. Temperature is the hotness or coldness of a substance, measured by one of the temperature scales. Heat is a form of energy that is measured in calories, with a calorie being the amount of heat that is necessary to raise 1 g of water by 1^0 C (Celsius). The specific heat is an inherent property of every substance. It is the amount of heat, in calories necessary to raise the temperature of 1 g of the substance by 1^0 C.

QUICK QUIZ

1.4
(a) False	(b) False	(c) True	(d) True
(e) False	(f) True	(g) True	(h) False
(i) True			

1.5
(a) False	(b) True	(c) True	(d) False
(e) True	(f) False		

1.8
(a) True	(b) False	(c) False	(d) False
(e) False	(f) False		

1.9
(a) False	(b) True	(c) False	(d) False
(e) False			

23

Learning in old age is writing on sand but
learning in youth is engraving on stone.
    - Arabian proverb

2

Atoms

CHAPTER OBJECTIVES

This chapter begins to present the fundamental aspects of chemistry. Whether or not you have been introduced to most of the terms and concepts previously in high school, you should very carefully master the material since it provides the real starting point and the basic grounding for subsequent chapters. You will also find that an understanding of the biochemistry associated with a living organism relies heavily on the basic concepts and principles of chemistry.

After studying the chapter and working the assigned exercises in the text and the study guide, you should be able to do the following.

1. Summarize and contrast the ideas of Zeno, Democritus, Dalton and Bohr regarding the basic makeup of matter.

2. Name and write out the chemical symbols for the first twenty elements (this is a minimum objective that should be extended).

3. Distinguish between a pure compound and a mixture.

4. State the four principles that serve as a basis for Dalton's atomic theory.

5. Characterize the elementary particles in the atom, including their charge, mass and location within the atom.

6. Given the number of protons and neutrons in an atom, determine the charge of the nucleus and the atomic number and mass number for the element.

7. Given the atomic number of an element, locate and identify the element on the Periodic Table.

8. Distinguish between an atom, isotope, a positive and negative ion and a molecule.

9. Explain the difference between a metal, metalloid, and a nonmetal and locate the general region where each class is located in the Periodic Table.

10. Write out the ground state electronic configuration for at least the first 40 elements..

11. Write out the ground state electronic configuration for positively and negatively charged ions of the first 20 elements.

12. Explain the trend in the ionization energies of the elements in a column (i.e., a group) and in a horizontal row on the Periodic Table.

13. List the two most abundant elements in the earth's crust and the four most important elements in the human body.

14. Define the important terms and comparisons in this chapter and give specific examples where appropriate.

IMPORTANT TERMS AND COMPARISONS

Atom, Ion and Molecule
Element
Periodic Table
Main-Group Elements
Representative Elements
Transition Metals and Inner Transition Metals
Isotopes
Law of Conservation of Mass
Ground State and Excited State
Law of Constant Composition
Principal Energy Levels (Shells) and Subshells
Unpaired and Paired Electrons
Inner and Outer Shell
Electronic Configuration
Monatomic, Diatomic and Polyatomic Elements
Valence Shell and Valence Electrons
Orbital Box Diagrams

Atomic Weight
Compound and Mixture
Group Numbers and Group Names
Alkali Metals
Metals, Metalloids and Nonmetals
Halogens and Noble Gases
Dalton's Atomic Theory
Quantized Energy Levels
Ionization Energy
s, p, and d Orbitals
Proton, Neutron and Electron
AMU
Nucleons and Nucleus
Mass Number and Atomic Number
Orbitals
Lewis Dot Structures

SELF-TEST QUESTIONS

MULTIPLE CHOICE. In the following exercises, select the correct answer from the choices listed. In some cases, two or more answers will be correct.

1. The concept of quantized energy levels in the atom was first proposed by:

   a. Zeno
   b. Bohr
   c. Lavoisier
   d. Democritus

2. Dalton's atomic theory includes the following statements:

   a. All atoms of the same element are identical to each other.
   b. All atoms are colorless.
   c. Atoms combine to form molecules. A molecule is a tightly bound group of atoms that acts as a unit.
   d. All matter is made up of very tiny indivisible particles called atoms.

3. The mass number of an atom that has 20 neutrons and a charge of +18 in the nucleus is:

   a. 20
   b. 38
   c. 56
   d. cannot determine

4. The atomic number of the atom in question 3 is:

   a. 20
   b. 38
   c. 18
   d. cannot determine

5. The ground state electronic configuration for the atom that has 6 protons and 6 neutrons in the nucleus is:

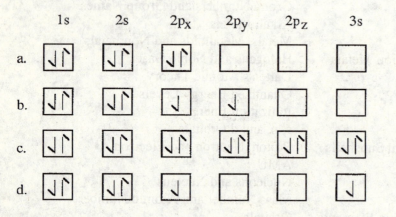

6. A possible excited state electronic configuration for the atom in question 5 is answer
   _____.

7. Which of the following elements is a metalloid?

   a. Hydrogen, H
   b. Germanium, Ge
   c. Silver, Ag
   d. Helium, He

26

8. Which of the following collection of elements are nonmetals?

    a. Na, K, Rb   c. O, S, Se
    b. F, Cl, Br   d. Cu, Ag, Au

9. The four most abundant elements in the earth's crust, listed in increasing order, are:

    a. Ca < Fe < O < Si   c. N > C > H > O
    b. Fe < Al < Si < O   d. Cl < O < N < P

10. The mass of a single atom of lead (Pb) is:

    a. 3.5 g   c. 3.5×10^{-22} g
    b. 3.5 ng   d. 3.5×10^{6} g

11. The four most abundant elements in the human body are:
    a. H, Li, C & N   c. C, N, S, & H
    b. C, N, O & H   d. Fe, C, N & Na

12. The four most common elements in the human body are considered:

    a. transition metals   c. representative elements
    b. metalloids   d. metals

13. The two most abundant elements in the crust of the earth are:

    a. Ag & Au   c. H & C
    b. O & Si   d. H & He

14. An example of periodicity can be seen in comparisons of:

    a. H, Li, Na & K   c. H, He, B & C
    b. C, N, Al & Si   d. Li, Be, C & N

15. The Lewis dot structure for sulfur is:

    a. S   c. S

    b. S   d. S

16. The size of a typical atom is about 10^{-10} m. The size of the nucleus of the atom is about:

    a. 10^{-10} m   c. 10^{-9} m
    b. 10^{-1} m   d. 10^{-15} m

<u>COMPLETION</u>. Write the word, phase or number in the blank that will complete the statement or answer the question.

1. The three elementary particles in an atom are the _____, _____ and _____, with only _____ of them being classified as residing in the nucleus (nucleons).

2. Of the alkali metals, Li, Na, K, Rb and Cs, the element, _____, requires the least energy to ionize an electron and to produce a positive ion.

3. The chemical symbols for the elements beryllium, arsenic, potassium and sodium are _____, _____, _____ and _____, respectively.

4. The atomic nucleus always possesses a _____ charge because the _____ resides in the nucleus.

5. Isotopes of an element have the same number of _____ , but different number of _____. Therefore, their _____ numbers are the same although their _____ numbers differ.

6. A positive ion is produced from an atom by _____ an _____ from the atom.

7. The halogens are in Group _____ and have _____ electrons in their outer shell.

8. Calcium, with atomic number 20, is in Group IIA. It is an example of a _____ element that has _____ electrons in its outer (valence) shell.

9. The ionization energy of elements in Group IA _____ going down the column, while the ionization energy going across a horizonal row of elements (from left to right) generally becomes _____.

10. The Periodic Table works because elements in the same _____ have the same _____ of _____ in the _____ shell.

11. In going down Group VIIA from flourine to astatine, the molecular weights progressively increase. Generally, as the molecular weights of similar compounds increase, the boiling points and melting points _____.

<u>QUICK QUIZ</u>

Indicate whether the statements are true or false. The number of the question refers to the section of the book that the question was taken from.

2.3 What are the Postulates of Dalton's Atomic Theory?
(a) Matter is divided into elements and pure substances.
(b) Matter is anything that has mass and volume (occupies space).
(c) A mixture is composed of two or more pure substances.
(d) An element is a pure substance.

28

(e) A heterogeneous mixture can be separated into pure substances, but a homogeneous mixture cannot.

(f) A compound consists of elements combined in a fixed ratio.

(g) A compound is a pure substance.

(h) All matter has mass.

(i) All of the 116 known elements occur naturally on Earth.

(j) The first six elements in the Periodic Table are the most important for human life.

(k) The combining ratio of a compound tells you how many atoms of each element are combined in the compound.

(l) The combining ratio of 1:2 in the compound CO_2 tells you that one gram of carbon is combined with two grams of oxygen.

2.4 What Are Atoms Made Of?

(a) A proton and an electron have the same mass but opposite charges.

(b) The mass of an electron is considerably smaller than that of a neutron.

(c) An atomic mass unit (amu) is a unit of mass.

(d) One amu is equal to 1 gram.

(e) The protons and neutrons of an atom are found in the nucleus.

(f) The electrons of an atom are found in the space surrounding the nucleus.

(g) All atoms of the same element have the same number of protons.

(h) All atoms of the same element have the same number of protons and neutrons.

(i) Electrons and protons have opposite charges.

(j) The size of an atom is approximately the size of its nucleus.

(k) Mass number is the sum of the numbers of protons and neutrons in the nucleus of an atom.

(l) For most atoms, the mass number is the same as the atomic number.

(m) The three isotopes of hydrogen (hydrogen-1, hydrogen-2, and hydrogen-3) differ only in the number of neutrons in the nucleus.

(n) Hydrogen-1 has one neutron in its nucleus, hydrogen-2 has two neutrons in its nucleus, and hydrogen-3 has three.

(o) All isotopes of an element have the same number of electrons.

(p) Most elements found on Earth are mixtures of isotopes.

(q) The atomic weight of an element given in the Periodic Table is the weighted average of the masses of its isotopes found on Earth.

(r) The atomic weights of most elements are whole numbers.

(s) Most of the mass of an atom is found in its nucleus.

(t) The density of a nucleus is its mass number expressed in grams.

2.5 What is the Periodic Table?

(a) Mendeleev discovered that, when elements are arranged in order of increasing atomic weight, certain sets of properties recur periodically.

(b) Main-group elements are those in the A columns of the Periodic Table.

(c) Nonmetals are found at the top of the Periodic Table, metalloids in the middle, and metals at the bottom.

(d) Among the 116 known elements, there are approximately equal numbers of metals and nonmetals.

(e) A horizontal row in the Periodic Table is called a group.

(f) The Group 1A elements are called the "alkali metals."

(g) The alkali metals react with water to give hydrogen gas and a metal hydroxide, MOH, where "M" is the metal.

(h) The halogens are all Group 7A elements.

(i) The boiling points of noble gases (Group 8A elements) increase in going from top to bottom of the column.

2.6 How Are the Electrons in an Atom Arranged?

(a) To say that "energy is quantized" means that only certain energy values are allowed.

(b) Bohr discovered that the energy of an electron in an atom is quantized.

(c) Electrons in atoms are confined to regions of space called "principal energy levels."

(d) Each principal energy level can hold a maximum of two electrons.

(e) An electron in a 1s orbital is held closer to the nucleus than an electron in a 2s orbital.

(f) An electron in a 2s orbital is harder to remove from an atom than an electron in a 1s orbital.

(g) An s orbital has the shape of a sphere, with the nucleus at the center of the sphere.

(h) Each 2p orbital has the shape of a dumbbell, with the nucleus at the midpoint of the dumbbell.

(i) The three 2p orbitals in an atom are aligned parallel to each other.

(j) An orbital is a region of space that can hold two electrons.

(k) The second shell contains one 2s orbital and three 2p orbitals.

(l) In the ground-state electron configuration of an atom, only the lowest-energy orbitals are occupied.

(m) A spinning electron behaves as a tiny bar magnet, with a north pole and a south pole.

(n) An orbital can hold a maximum of two electrons with their spins paired (oriented in opposite directions).

(o) Paired electron spins means that the two electrons are aligned with their spins north pole to north pole and south pole to south pole.

(p) An orbital box diagram puts all of the electrons of an atom in one box with their spins aligned.

(q) An orbital box diagram of a carbon atom shows two unpaired electrons.

(r) A Lewis dot structure shows only the electrons in the valence shell of an atom of the element.

(s) A characteristic of Group 1A elements is that each has one unpaired electron in its outermost occupied (valence) shell.

(t) A characteristic of Group 6A elements is that each has six unpaired electrons in its valence shell.

2.7 How Are Electron Configuration and Position in the Periodic Table Related?

(a) Elements in the same column of the Periodic Table have the same outer-shell electron configuration.

(b) All Group 1A elements have one electron in their valence shell.

(c) All Group 6A elements have six electrons in their valence shell.

(d) All Group 8A elements have eight electrons in their valence shell.

(e) Period 1 of the Periodic Table has one element, period 2 has two elements, period 3 has three elements, and so forth.

(f) Period 2 results from filling the 2s and 2p orbitals and, therefore, there are eight elements in period 2.

30

(g) Period 3 results from filling the $3s$, $3p$, and $3d$ orbitals and, therefore, there are nine elements in period 3.

(h) The main-group elements are s block and p block elements.

2.8 What Is A Periodic Property?

(a) Ionization energy is the energy required to remove the most loosely held electron from an atom in the gas phase.

(b) When an atom loses an electron, it becomes a positively charged ion.

(c) Ionization energy is a periodic property because ground-state electron configuration is a periodic property.

(d) Ionization energy generally increases in going from left to right across a period of the Periodic Table.

(e) Ionization energy generally increases in going from top to bottom within a column in the Periodic Table.

(f) The sign of an ionization energy is always positive; the process is always endothermic.

ANSWERS TO SELF-TEST QUESTIONS

MULTIPLE CHOICE

1.	b	9.	b
2.	a, c, d	10.	c
3.	b	11.	b
4.	c	12.	c
5.	b	13.	b
6.	a or d	14.	a
7.	b	15.	a
8.	b, c	16.	d

COMPLETION

1. proton, neutron, electron, 2
2. Cesium (Cs)
3. Be, As, K and Na
4. positive, proton
5. protons and electrons, neutrons, atomic, mass
6. pulling off or ionizing, electron from
7. VII, 7
8. representative or a metallic, 2
9. decreases, greater
10. group or column, number of electrons, outer
11. increase

31

QUICK QUIZ

2.3

(a) False	(b) True	(c) True	(d) True
(e) False	(f) True	(g) True	(h) True
(i) False	(j) False	(k) True	(l) False

2.4

(a) False	(b) True	(c) True	(d) False
(e) True	(f) True	(g) True	(h) False
(i) True	(j) False	(k) True	(l) False
(m) True	(n) False	(o) True	(p) True
(q) True	(r) False	(s) True	(t) False

2.5

(a) True	(b) True	(c) False	(d) False
(e) False	(f) True	(g) True	(h) True
(i) True			

2.6

(a) True	(b) True	(c) True	(d) False
(e) True	(f) False	(g) True	(h) True
(i) False	(j) True	(k) True	(l) True
(m) True	(n) True	(o) False	(p) False
(q) True	(r) True	(s) True	(t) False

2.7

(a) True	(b) True	(c) True	(d) True
(e) False	(f) True	(g) False	(h) True

2.8

(a) True	(b) True	(c) True	(d) True
(e) False	(f) True		

"I believe the chemical bond is not so
simple as some people seem to think."
- Robert S. Mulliken

3

Chemical Bonds

CHAPTER OBJECTIVES

Although much remains to be learned about the atom, the titles of this and all subsequent chapters reveal that a great deal of attention is devoted to understanding the chemistry of ions and molecules. In either case, the outer **electronic configuration of the atom is altered** by the gain or loss of an electron to form a charged particle, or alternatively, electrons are shared in a common covalent bond between two atoms. As a result, the chemical and physical properties of these chemical species are different than the atoms from which they were derived. Also, in the case of complex ions or molecules, the species take on specific three-dimensional shapes, unlike that of the beautifully simple spherical shape of an atom. Structures for these molecules, called **Lewis structures**, are used to describe the electronic arrangement of the **valence electrons**. In some cases, contributing **resonance structures** are needed to adequately describe the molecule. The molecular shapes can be predicted from the **valence-shell electron-pair repulsion (VSEPR)** model.

After you have studied this chapter and worked the exercises in the text and study guide, you should be able to do the following.

1. Write the equations for the formation of a cation or anion from the parent atom.

2. Using the octet rule as a guide, predict why certain ions are stable, while others are nonexistent.

3. From a knowledge of the charge on the positive and negative ions and using the principle of electrical neutrality, predict the formulas for a variety of ionic compounds.

4. Distinguish between an atom, an ion and a molecule.

5. Describe the characteristic difference between an ionic bond and a covalent bond.

6. Characterize a single, double and triple bond and give specific examples of compounds that contain these bond types.

7. Write out the molecular formula for a variety of compounds, in addition to drawing the structural formula and Lewis structure for each compound.

8. State the VSEPR model. Given the number of valence shell electron pairs of the central atom of a molecule and the nature of bonded atoms, predict the shape of the molecule and the size of the angle between the bonds in each. Give specific examples of molecules with each of the major molecular shapes.

9. Define electronegativity and explain the trend in electronegativity values in the Periodic Table as shown in Table 4.5 (text).

10. Draw the structural formulas for some di-, tri- and tetraatomic molecules. Given the electronegativity value for each atom, indicate whether the individual **bonds** and whether the **molecules** themselves are polar or nonpolar.

11. Using the general rules which relate the electronegativity differences of the atoms in a bond to the bond type, predict whether the bonds in a variety of diatomic molecules are ionic, polar-covalent or nonpolar-covalent.

12. Write out the name, together with the molecular and structural formulas for some common cations and anions (utilize Table 3.1- 3.4, text) and some medicinally important polyatomic ions (Box 3B, text).

13. Write out the contributing (resonance) structures for molecules that cannot be adequately described by a single Lewis structure. Show all atoms, electrons and the charges.

14. Write out the names and chemical formulas for some of the inorganic compounds that are found in the chapter.

15. Define the important terms and comparisons in this chapter and give specific examples where appropriate.

IMPORTANT TERMS AND COMPARISONS

Ion, Cation, Anion	Molecular and Structural Formula
Ionic and Covalent Bonds	Lewis Structures
Three-Dimensional Molecular Structure	Electrical Neutrality
Monatomic, Diatomic and Polyatomic Ions	Hydroxyapatite
VSEPR Theory	Bond Angle

Valence Electrons

Trigonal Planar, Pyramidal and Tetrahedral Shapes

Electronegativity, Polar Bonds

Single, Double and Triple Bond

Polar Covalent and Non-Polar Covalent Bonds

Polar Covalent and Non-Polar Covalent Molecules

Linear and Angular Molecules

Bonding and Nonbonding Electrons

Dipoles and Partial Charges, δ^+, δ^-

Resonance Structures, Resonance Hybrid

Exceptions to the Octet Rule

SELF-TEST QUESTIONS

MULTIPLE CHOICE. In the following exercises, select the correct answer from the choices listed. In some cases, two or more answers will be correct.

1. The formula for cesium phosphate is:

    a. $Cs_3(PO_4)$
    b. $Cs_2(P_4)$

    c. $Cs(PO_4)_3$
    d. $CsPO_4$

2. The Lewis structure for the ozone molecule, O_3, is:

    a.

    b.

    c.

    d.

3. Draw the Lewis structure for H_2CO. Indicate the number of electrons and the types of electron pairs that are around the oxygen atom.

    a. 3 bonding pairs and 1 non-bonding pair
    b. 2 bonding pairs and 2 non-bonding pairs

    c. 4 non-bonding pairs
    d. 4 bonding pairs

4. Which of the following ions is stable?

    a. F^+
    b. S^{2-}
    c. Ca^+

    d. O^{2+}
    e. I^-
    f. N^{3-}

5. A particular atom has the following characteristics. The atomic number is 12 and it forms cations with a +2 charge. This element is:

    a.   Na   c.   S
    b.   Al   d.   Mg

6. The bond that is most polar is:

    a.   O-H   c.   B-H
    b.   S-H   d.   F-H

7. The order of decreasing difference in polarity in the bonds in question 6 is:

    a.   H-F > O-H > S-H > B-H
    b.   H-F > S-H > B-H > O-H
    c.   O-H > H-F > S-H > B-H
    d.   S-H > O-H > H-F > B-H

8. Which of the following molecules have a correct Lewis structure?

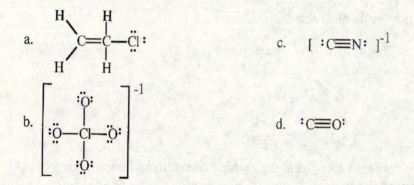

9. Indicate the following compound(s) that contains a covalent bond.

    a. $CaCl_2$ c. SO_2
    b.   ClBr   d.   KCl

10. Indicate which pair of atoms has the greatest electronegativity difference.

    a.   H & C   c.   F & Cl
    b.   H & O   d.   C & N

36

11. Predict the molecular formula for the compound that contains calcium and phosphate.

    a. $Ca(PO_4)$ c. $Ca_2(PO_4)_3$
    b. $Ca_3(PO_4)_2$ d. $Ca(PO_4)_2$

12. Indicate which of the following pairs of compounds is not named correctly.

a.	$CuCl$ & $CuCl_2$	cuprous chloride and cupric chloride
b.	$FeBr_3$ and $FeBr_2$	iron tribromide & iron dibromide
c.	SnO and SnO_2	stannous oxide & stannic oxide
d.	NO and N_2O	nitrogen oxide & dinitrogen oxide

13. Indicate which of the following statements about the relative electronegativity of two atoms is not correct.

    a.   Br is greater than As.
    b.   Li is greater than K
    c.   B is less than Al
    d.   Na is less than Mg

14. The molecular formula for hydroxyapatite is $Ca_5(PO_4)_x(OH)$. Indicate how many phosphate units are in this formula if there is only one OH.

    a.   1   c.   3
    b.   2   d.   4

15. Which of the following compounds must be represented by two or more Lewis structures and is represented by a resonance hybrid?

    a. H_2O c. O_3
    b. CO_2 d. CO_3^{-2}

COMPLETION. Write the word, phrase or number in the blank that will complete the statement or answer the question.

1. Based on the VSEPR theory, determine the molecular shape of each molecule.

    a. AlF_4^- _____

    b. C_2HCl _____

    c. H_3O^+ _____

    d. PCl_3 _____

e. BI_3 _____

f. I_3^- _____
       (The central I does not obey
       the octet rule.)

2.   Classify whether the bond between the two elements shown is non-polar, covalent, polar covalent or ionic.

    a.   C, C   _____

    b.   Li, F   _____

    c.   Na, O   _____

    d.   H, I   _____

    e.   O, S   _____

3.   Name these compounds or ions.

    a. CN^- _____

    b. $(NH_4)_2SO_4$ _____

    c. NO_2^- _____

    d. K_2O _____

    e. $MgCl_2$ _____

    f. $Li(CH_3COO)$ _____

4.   Indicate the bond angles in these molecules.

    a. CO_2 _____

    b. CCl_4 _____

    c. H_2O _____

    d. BCl_3 _____

    e. C_2H_2 _____

5. Write out the chemical formulas for these compounds.

    a.   sodium bicarbonate   _____

    b.   rubidium sulfate   _____

    c.   magnesium permanganate   _____

    d.   ammonium oxide   _____

6. The formation of a positively charged _____ from an atom requires the _____ of an electron, while the formation of a negatively charged _____ requires the _____ of an electron.

7. A _____ bond involves the _____ of one or more electron pairs, while an _____ bond results from an electrostatic attraction between _____ particles.

8. Although the C^{4+} cation has a complete outer shell, it does not exist because such a ____ _____ is unstable.

9. In solid NaCl, each Na^+ has _____ Cl^- ions as _____ _____.

10. In resonance structures for a molecule, one finds that the _____ and _____electrons and (in some cases) the _____ can be moved from one place to another, but the _____ cannot be moved.

11. Name these compounds.

    a. $NiCl_2$ _____

    b. $Cr(NO_3)_3$ _____

    c. $NaH(SO_4)$ _____

    d. MgF_2 _____

12. Write the chemical formula for each of these compounds.

    a.   copper(II) iodide   _____

    b.   sodium oxide   _____

    c.   aluminum oxide   _____

13. From VSEPR theory, the predicted bond angle for OF2 would be about _____ degrees, while that for CO2 would be _____.

14. One finds that cations formed from the Groups (two groups) _____ & _____ atoms exhibit only one characteristic charge, while metals classified as _____ metals can exist in more than one positive charge.

QUICK QUIZ
Indicate whether the statements are true or false. The number of the question refers to the section of the book that the question was taken from.

3.2 What is the Octet Rule?
(a) The octet rule refers to the chemical bonding patterns of the first eight elements of the Periodic Table.
(b) The octet rule refers to the tendency of certain elements to react in such a way that they achieve an outer shell of eight valence electrons.
(c) In gaining electrons, an atom becomes a positively charged ion called a cation.
(d) When an atom forms an ion, only the number of valence electrons changes; the number of protons and neutrons in the nucleus does not change.
(e) In forming ions, Group 2A elements typically lose two electrons to become cations with a charge of +2.
(f) In forming an ion, a sodium atom $(1s^2 2s^2 2p^6 3s^1)$ completes its valence shell by adding one electron to fill its 3s shell $(1s^2 2s^2 2p^6 3s^2)$.
(g) Group 6A elements typically react by accepting two electrons to become anions with a charge of -2.
(h) With the exception of hydrogen, the octet rule applies to all elements in periods 1,2, and 3.
(i) Atoms and the ions derived from them have very similar physical and chemical properties.

3.3 How Do We Name Cations and Anions?
(a) For Group 1A and Group 2A elements, the name of the ion each forms is simply the name of the element; for example, Mg^{2+} is named magnesium ion.
(b) H^+ is called the hydronium ion.
(c) The nucleus of H^+ consists of one proton and one neutron.
(d) Many transition and inner transition elements form more than one positively charged ion.
(e) In naming metal cations with two different charges, the suffix -ous refers to the ion with a charge of +1, and -ic refers to the ion with a charge of +2.
(f) Fe^{3+} may be named either iron(III) ion or ferric ion.
(g) The anion derived from a bromine atom is named bromine ion.
(h) The anion derived from an oxygen atom is named oxide ion.
(i) HCO_3^- is named hydrogen carbonate ion.
(j) The prefix bi- in the name "bicarbonate" ion indicates that this ion has a charge of -2.
(k) The hydrogen phosphate ion has a charge of +1 and the dihydrogen phosphate ion has a charge of +2.
(l) The phosphate ion is PO_3^{4-}.
(m) The nitrite ion is NO_2^- and the nitrate ion is NO_3^-.

40

(n) The carbonate ion is CO_3^{2-} and the hydrogen carbonate ion is HCO_3^-.

3.4 What Are the Two Major Types of Chemical Bonds?

(a) According to the Lewis model of bonding, atoms bond together in such a way that each atom participating in the bond has an outer-shell electron configuration matching that of the noble gas nearest to it in atomic number.

(b) Atoms that lose electrons to achieve a filled valence shell become cations and form ionic bonds.

(c) Atoms that gain electrons to achieve filled valence shells become anions and form ionic bonds.

(d) Atoms that share electrons to achieve filled valence shells form covalent bonds.

(e) Ionic bonds tend to form between elements on the left side of the Periodic Table and covalent bonds tend to form between elements on the right side of the Periodic Table.

(f) Ionic bonds tend to form between a metal and a nonmetal.

(g) When two nonmetals combine, the bond between them is usually covalent.

(h) Electronegativity is a measure of an atom's attraction for the electrons it shares in a chemical bond with another atom.

(i) Electronegativity generally increases with atomic number.

(j) Electronegativity generally increases with atomic weight.

(k) Electronegativity is a periodic property.

(l) Fluorine, in the upper-right corner of the Periodic Table, is the most electronegative element; hydrogen, in the upper-left corner is, the least electronegative element.

(m) Electronegativity depends on both the nuclear charge and on the distance between the valence electrons and the nucleus.

(n) Electronegativity generally increases from left to right across a period of the Periodic Table.

(o) Electronegativity generally increases from top to bottom in a column of the Periodic Table.

3.5 What Is An Ionic Bond?

(a) An ionic bond is formed by the combination of positive and negative ions.

(b) An ionic bond between two atoms forms by the transfer of one or more valence electrons from the atom of higher electronegativity to the atom of lower electronegativity.

(c) As a rough guideline, we say that an ionic bond will form if the difference in electronegativity between two atoms is approximately 1.9 or greater.

(d) In forming NaCl from sodium and chlorine atoms, one electron is transferred from the valence shell of sodium to the valence shell of chlorine.

(e) The formula of sodium sulfide is Na_2S.

(f) The formula of calcium hydroxide is CaOH.

(g) The formula of aluminum sulfide is AlS.

(h) The formula of iron(III) oxide is Fe_3O_2.

(i) Barium ion is Ba^{2+} and oxide ion is O^{2-} and, therefore, the formula of barium oxide is Ba_2O_2.

3.6 How Do We Name Ionic Compounds?

(a) The name of a binary ionic compound consists of the name of the positive ion followed by the name of the negative ion.

41

(b) In naming binary ionic compounds, it is necessary to show the number of each ion present in the compound.

(c) The formula of aluminum oxide is Al_2O_3.

(d) Both copper(II) oxide and cupric oxide are acceptable names for CuO.

(e) The systematic name for Fe_2O_3 is iron(II) oxide.

(f) The systematic name for $FeCO_3$ is iron carbonate.

(g) The systematic name for NaH_2PO_4 is sodium dihydrogen phosphate.

(h) The systematic name for K_2HPO_4 is dipotassium hydrogen phosphate.

(i) The systematic name for Na_2O is sodium oxide.

(j) The systematic name for PCl_3 is potassium chloride.

(k) The formula of ammonium carbonate is NH_4CO_3.

3.7 What Is a Covalent Bond?

(a) A covalent bond is formed between two atoms whose difference in electronegativity is less than 1.9.

(b) If the difference in electronegativity between two atoms is zero (they have identical electronegativity values), then the two atoms will not form a covalent bond.

(c) A covalent bond formed by sharing two electrons is called a double bond.

(d) In the hydrogen molecule (H_2), the shared pair of electrons completes the valence shell of each hydrogen.

(e) In the molecule CH_4, each hydrogen has an electron configuration like that of helium, and carbon has an electron configuration like that of neon.

(f) In a polar covalent bond, the more electronegative atom has a partial negative charge ($\delta-$) and the less electronegative atom has a partial positive charge ($\delta+$).

(g) These bonds are arranged in order of increasing polarity C-H < N-H < O-H.

(h) These bonds are arranged in order of increasing polarity H-F < H-Cl < H-Br.

(i) A polar bond has a dipole with the negative end at the more electronegative atom.

(j) In a single bond, two atoms share one pair of electrons, in a double bond, they share two pairs of electrons; and in a triple bond, they share three pairs of electrons.

(k) The Lewis structure for ethane, C_2H_6, must show eight valence electrons.

(l) The Lewis structure for formaldehyde, CH_2O, must show 12 valence electrons.

(m) The Lewis structure for the ammonium ion, NH_4^+, must show nine valence electrons.

(n) Atoms of period 3 elements can hold more than eight electrons in their valence shells.

3.8 How Do We Name Binary Covalent Compounds?

(a) A binary covalent compound contains two kinds of atoms.

(b) The two atoms in a binary covalent compound are named in this order: first the more electronegative element and then the less electronegative element.

(c) The name for SF_2 is sulfur difluoride.

(d) The name of CO_2 is carbon dioxide.

(e) The name of CO is carbon oxide.

(f) The name of HBr(g) is hydrogen bromide.

(g) The name of CCl_4 is carbon tetrachloride.

3.9 What is Resonance?

(a) Two chemical structures in equilibrium with each other are contributing structures to the resonance hybrid.

(b) A double-headed arrow is used to indicate that each structure is a contributing structure to the resonance hybrid.

(c) The resonance hybrid is more stable than any of the contributing structures.

(d) Although the resonance hybrid must obey the rules of covalent bonding, the contributing structures do not have this limitation.

3.10 How Do We Predict Bond Angles in Covalent Molecules?

(a) The letters VSEPR stand for valence-shell electron-pair repulsion.

(b) In predicting bond angles about a central atom in a covalent molecule, the VSEPR model considers only shared electron pairs (electron pairs involved in forming covalent bonds).

(c) The VSEPR model treats the two electron pairs of a double bond as one region of electron density, and the three electron pairs of a triple bond as one region of electron density.

(d) In carbon dioxide, O=C=O, carbon is surrounded by four pairs of electrons and the VSEPR model predicts 109.5° for the O-C-O bond angle.

(e) For a central atom surrounded by three regions of electron density, the VSEPR model predicts bond angles of 120°.

(f) The geometry about a carbon atom surrounded by three regions of electron density is described as trigonal planar.

(g) For a central atom surrounded by four regions of electron density, the VSEPR model predicts bond angles of 360°/4 = 90°.

(h) For the ammonia molecule, NH_3, the VSEPR model predicts H-N-H bond angles of about 109.5°.

(i) For the ammonium ion, NH_4^+, the VSEPR model predicts H-N-H bond angles of 109.5°.

(j) The VSEPR model applies equally well to covalent compounds of carbon, nitrogen, and oxygen.

(k) In water, H-O-H, the oxygen atom forms covalent bonds to two other atoms and, therefore, the VSEPR model predicts an H-O-H bond angle of 180°.

(l) If you fail to consider unshared pairs of valence electrons when you use the VSEPR model, you will arrive at an incorrect prediction.

(m) Given the assumptions of the VSEPR model, the only bond angles it predicts for compounds of carbon, nitrogen, and oxygen are 109.5°, 120°, and 180°.

3.11 How Do We Determine if a Molecule Is Polar?

(a) To predict whether a covalent molecule is polar or nonpolar, you must know both the polarity of each bond and the geometry (shape) of the molecule.

(b) A molecule may have two or more polar bonds and still be nonpolar.

(c) All molecules with polar bonds are polar.

(d) If water were a linear molecule with an H-O-H bond angle of 180°, water would be a nonpolar molecule.

(e) H_2O and NH_3 are polar molecules, but CH_4 is nonpolar.

(f) In methanol, CH_3OH, the O-H bond is more polar than the C-O bond.

(g) Dichloromethane, CH_2Cl_2, is polar but tetrachloromethane, CCl_4, is nonpolar.

(h) Ethanol, CH_3CH_2OH, the alcohol of alcoholic beverages, has polar bonds, a net dipole, and is a polar molecule.

ANSWERS TO SELF-TEST QUESTIONS

MULTIPLE CHOICE

1. a
2. a (note that 18 valence electrons are involved and the octet rule must be obeyed)
3. b
4. b, e, f
5. d
6. d
7. a
8. b, c, d
9. b, c
10. b
11. c
12. b
13. b
14. c
15. c, d

COMPLETION

1.
   a. tetrahedral
   b. linear
   c. pyramidal
   d. pyramidal
   e. trigonal planar
   f. linear

2.
   a. nonpolar covalent, $\Delta E = 0$
   b. ionic, $\Delta E = 3.0$
   c. ionic, $\Delta E = 2.4$
   d. nonpolar covalent, $\Delta E = 0.4$
   e. polar covalent, $\Delta E = 1.0$

3.
   a. cyanide anion (or ion)
   b. ammonium sulfate
   c. nitrite anion (or ion)
   d. potassium oxide
   e. magnesium chloride
   f. lithium acetate

4.
   a. 180°
   b. 109.5°
   c. 105°
   d. 120°
   e. 180°

5.
   a. $Na(HCO_3)$
   b. $Rb_2(SO_4)$
   c. $Mg(MnO_4)_2$
   d. $(NH_4)_2O$

44

6. cation, removal, anion, addition
7. covalent, sharing, ionic, charged
8. large, positive charge
9. 6, nearest neighbors
10. bonding, nonbonding, charge, atoms
11. a. nickel(II) chloride
    b. chromium(III) nitrate
    c. sodium hydrogen sulfate
    d. magnesium fluoride
12. a. CuI_2
    b. Na_2O
    c. Al_2O_3
13. 109^0, 180^0
14. IA and IIA, transition

QUICK QUIZ

3.2

(a) False	(b) True	(c) False	(d) True
(e) True	(f) False	(g) True	(h) False
(i) False			

3.3

(a) True	(b) False	(c) False	(d) True
(e) False	(f) True	(g) False	(h) True
(i) True	(j) False	(k) False	(l) False
(m) True	(n) True		

3.4

(a) True	(b) True	(c) True	(d) True
(e) False	(f) True	(g) True	(h) True
(i) False	(j) False	(k) True	(l) False
(m) True	(n) True	(o) False	

3.5

(a) True	(b) False	(c) True	(d) True
(e) True	(f) False	(g) False	(h) False
(i) False			

3.6

(a) True	(b) False	(c) True	(d) True
(e) False	(f) False	(g) True	(h) False
(i) True	(j) False	(k) False	

3.7

(a) True	(b) False	(c) False	(d) True
(e) True	(f) True	(g) True	(h) False
(i) True	(j) True	(k) False	(l) True
(m) False	(n) True		

3.8

(a) True	(b) False	(c) True	(d) True
(e) False	(f) True	(g) True	

3.9

(a) False	(b) True	(c) True	(d) False

3.10

(a) True	(b) False	(c) True	(d) False
(e) True	(f) True	(g) False	(h) True
(i) True	(j) True	(k) False	(l) True
(m) True			

3.11

(a) True	(b) True	(c) False	(d) True
(e) True	(f) True	(g) True	(h) True

....chemical transformations....are
responsible for corrosion and decay,
for development, growth and life.
- C.N. Hinshelwood
The Kinetics of Chemical Change, 1940

4

Chemical Reactions

CHAPTER OBJECTIVES

Chemistry is a quantitative science. This means that it is possible to determine the exact quantity of a substance - for example, the volume, density, mass or number of molecules of water in a raindrop. Importantly, one can relate **macroscopic** quantities, such as the mass of a substance, to **microscopic** quantities, such as the number of atoms or molecules in a given quantity of the substance. In addition, exact relationships can be represented by a balanced chemical equation that describes the reaction that is taking place. For example, we all realize that it requires gasoline to run an automobile. This process of combustion of the gasoline can be described quantitatively by a balanced chemical equation. This chapter presents these concepts, relationships and the procedures that provide the basis for some of the quantitative aspects of chemistry.

After you have studied the chapter and worked the exercises in the text and study guide, you should be able to do the following.

1. Calculate the formula weight of any ionic or molecular compound from its molecular formula.

2. Calculate the number of moles in a given mass of a compound.

3. Using Avogadro's number of 6.02×10^{23}, determine the number of molecules in a given mass of a compound.

4. Readily interconvert between the mass of a substance and the number of moles and molecules of the compound.

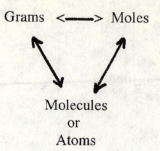

Grams <——> Moles

Molecules
or
Atoms

5. Using the atomic weights, determine the relative mass of pairs of atoms on the Periodic Table.

6. Balance simple chemical equations.

7. From a balanced chemical equation, carry out calculations to determine the mass and molar relationship between the reactants and products.

8. Using the balanced equation and given the actual quantities of the reactants used up and the product formed in a reaction, calculate the theoretical and percentage yield of product.

9. Write out the complete, balanced equations for the reaction of various ionic compounds in an aqueous solution. Point out the spectator ions and then rewrite the net ionic equation.

10. Write out the general reaction scheme for a combination and a decomposition reaction and then give a specific example of each.

11. Be able to recognize and balance a redox reaction. Indicate the species that are oxidized and reduced and the species which are the oxidizing and the reducing agent, respectively.

12. Explain how a voltaic cell works.

13. Given balanced equations, including the heat gained or lost in the reaction, generally classify the reaction type and indicate whether the reaction is exothermic or endothermic.

14. Define the important terms and comparisons in this chapter and give specific examples where appropriate.

IMPORTANT TERMS AND COMPARISONS

Chemical Reaction
Reactant and Product
Formula Weight and Molecular Weight
AMU
Avogadro's Number and Mole
Factor-Label Method
Conversion Factor

Symbols, ↑, ↓
Oxidation and Reduction (Redox)
Oxidizing and Reducing Agent
Combustion
Limiting Reagent
Respiration
Rusting

Unbalanced and Balanced Equation

Coefficient and Subscript

Conservation of Mass

Actual Yield, Theoretical Yield and Percent Yield

Calorie

Dissociation of Ionic Compounds

Spectator Ions

Combination and Decomposition Reactions

Voltaic Cell

Metabolism

Heat of Reaction and Heat of Combustion

Exothermic and Endothermic Reactions

Side Reactions

Aqueous (Solution)

Soluble and Insoluble Compounds

Net Ionic Equation

Antiseptic and Disinfectant

Cathode and Anode

FOCUSED REVIEW

To become comfortable and efficient with the mathematical manipulations of the new material, work the following exercises before going onto the SELF-TEST QUESTIONS.

<u>Exercise I</u>. <u>Calculation of Formula Weights</u>. From the molecular formulas, calculate the formula weight of each compound. Round the answer off to the first decimal place.

a. $Mg(ClO_4)_2$ _____

b. Na_2O _____

c. $Al(CH_3CO_2)_3$ _____

d. $FeBr_3$ _____

e. O_3 _____

f. $C_{18}H_{36}O_2$ _____

g. $C_3H_7NO_2$ _____

h. $Pt(NH_3)_2Cl_2$ _____

<u>Exercise II</u>.   <u>Interconversion of Grams, Moles and Molecules</u>. The most important interrelationship in this chapter is that between grams, moles and molecules and is represented below.

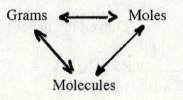

In converting one of these quantities to another, a conversion unit is available and is used "as is" or inverted, depending on the direction of the conversion (i.e., whether converting from grams to moles or from moles to grams).

Using H_2O as an example, it will be (or is) obvious that 1 gram of water is equivalent to the following.

$$1.0 \text{ g } H_2O \leftrightarrow 5.5 \text{x} 10^{-2} \text{ moles } H_2O \leftrightarrow 3.3 \text{x} 10^{22} \text{ molecules } H_2O$$

To accomplish these interconversions, the quantity is multiplied by the appropriate conversion factor, making use of the factor-label method. The conversion factor in each of these examples is underlined.

For example:

Given: 1.0 g H_2O Convert to: moles H_2O

$$\left(\frac{1.0 \text{ g } H_2O}{1} \right) \left(\frac{1 \text{ mole } H_2O}{18 \text{ g } H_2O} \right) = 5.5 \text{x} 10^{-2} \text{ moles } H_2O$$

Given: 1.0 g H_2O Convert to: molecules H_2O

$$\left(\frac{1.0 \text{ g } H_2O}{1} \right) \left(\frac{1 \text{ mole } H_2O}{18 \text{ g } H_2O} \right) \left(\frac{6.02 \text{x} 10^{23} \text{ molecules}}{1 \text{ mole } H_2O} \right) = 3.3 \text{x} 10^{22} \text{ molecules } H_2O$$

Given: $5.5 \text{x} 10^{-2}$ moles H_2O Convert to: g H_2O

$$\left(\frac{5.5 \text{x} 10^{-2} \text{ moles } H_2O}{1} \right) \left(\frac{18 \text{ g}}{1 \text{ mole } H_2O} \right) = 1.0 \text{ g } H_2O$$

Given: $5.5 \text{x} 10^{-2}$ moles H_2O Convert to: molecules H_2O

$$\left(\frac{5.5 \text{x} 10^{-2} \text{ moles } H_2O}{1} \right) \left(\frac{6.02 \text{x} 10^{23} \text{ molecules}}{1 \text{ mole } H_2O} \right) = 3.3 \text{x} 10^{22} \text{ molecules } H_2O$$

Given: $3.3 \text{x} 10^{22}$ molecules H_2O Convert to: moles H_2O

$$\left(\frac{3.3 \text{x} 10^{22} \text{ molecules } H_2O}{1} \right) \left(\frac{1 \text{ mole } H_2O}{6.02 \text{ x } 10^{23} \text{ molecules}} \right) = 5.5 \text{x} 10^{-2} \text{ moles } H_2O$$

Given: 3.3×10^{22} molecules H_2O Convert to: g H_2O

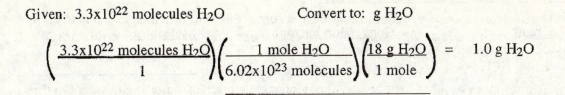

$$\left(\frac{3.3 \times 10^{22} \text{ molecules } H_2O}{1} \right) \left(\frac{1 \text{ mole } H_2O}{6.02 \times 10^{23} \text{ molecules}} \right) \left(\frac{18 \text{ g } H_2O}{1 \text{ mole}} \right) = 1.0 \text{ g } H_2O$$

As noted in the text, the conversion from grams to moles or visa versa requires only one step, while conversion between grams and molecules requires 2 steps or 2 conversion factors. The importance of using the factor-label method is brought out in these calculations as well.

Carry out the following conversions, indicating the conversion unit used in the factor-label analysis. In addition, the answer should have the correct number of significant figures.

A.   GRAMS ↔ MOLES CONVERSION

Grams	Conversion Factor	Moles
(i) 4.0 g HF	_____	_____
(ii) 33 g KOH	_____	_____
(iii) 0.020 g AlF_3	_____	_____
(iv) _____	_____	0.30 moles PCl_3
(v) _____	_____	1.2 moles GeH_4
(vi) _____	_____	0.01 moles H_2CO_2

B.   MOLES ↔ MOLECULES CONVERSION

Grams	Conversion Factor	Molecules
(i) 0.34 moles H_2	_____	_____
(ii) 11.3 moles Na_2SO_4	_____	_____
(iii) 3.4×10^{-24} moles NH_3	_____	_____
(iv) _____	_____	1 molecule of I_2
(v) _____	_____	6×10^{20} molecules O_3
(vi) _____	_____	4×10^{6} molecules $HClO_2$

C. GRAMS ↔ MOLECULES CONVERSION

Grams	Conversion Factor	Molecules
(i) 0.14 g NCl_3	_____	_____
(ii) 4.0 mg S_8	_____	_____
(iii) 7.4 ug HI	_____	_____
(iv) _____	_____	4×10^{27} molecules $SeCl_2$
(v) _____	_____	6×10^{12} molecules $C_2H_4O_2$
(vi) _____	_____	2×10^2 molecules GeI_4

Exercise III.   Stoichiometry

In determining the weight relationships in chemical reactions, the key points are these:

1.   The equation must be **balanced.**

2.   **Molecules** react with **molecules. Moles** react with **moles.** The coefficients in the balanced equation dictate exactly how many of each of the reactants will react to form exactly how many product molecules. Therefore, the number of **moles** of the known reactant or product must be calculated first, if it is not given.

3.   The coefficients in the balanced equation indicate the relative number of moles of each reactant and product. Therefore, multiply the moles of the known compound by the coefficient of the compound whose weight is unknown divided by the coefficient of the compound of known weight (or moles).

4.   Convert the number of moles back to weight, by use of the appropriate conversion factor.

Consider the following equation. From the quantity given, determine the number of moles of the other quantities of interest.

$$C_7H_{16} + 11 O_2 \rightarrow 7 CO_2 + 8 H_2O$$

A.   Given: 0.1 mole C_7H_{16}

moles O_2	moles CO_2	moles H_2O
_____	_____	_____

52

B.   Given: 0.1 mole O_2

moles C_7H_{16}	moles CO_2	moles H_2O
_____	_____	_____

C.   Given 0.10 mole CO_2 produced

moles C_7H_{16}	moles O_2	moles H_2O
_____	_____	_____

Exercise IV.   Limiting Reagent.   In most "real-life" situations, there are reactions or relationships that involve a limitation or a restriction that there is a limited amount of one of the reactants and therefore, the amount of product formed will be limited.

A few simple (non-chemistry) examples may help.

A football program has all the football gear (shoulder pads, helmets, etc) to have a team of 100 players. However, they have only 75 shirts with the player's numbers on them. As a result, they can have only 75 players on the team. The number of shirts is limiting and the shirt is the limiting "reagent".

A chemistry course is taught in a classroom that can accommodate 150 students. However, the required laboratory can only accommodate 50 students. The enrollment in the course is limited to 50. The limiting "reagent" is the number of students that can be accommodated in the laboratory.

Let's try a chemical example.

$$1\ Na \quad + \quad 1/2\ Cl_2 \quad = \quad 1\ NaCl$$

	1 Na	1/2 Cl₂	1 NaCl
Available reagent (g)	22.9 g	200 g	------
Available reagents (moles)	1 mole	2.82 moles	------
Used reagent	1 mole	0.5 moles	------
Final amounts in the reaction	-----	2.32 moles	1 mole

An experimenter carries out the reaction with 22.9 g of Na metal and 200 g of Cl_2 gas. Calculations indicate (verify this) that there is 1 mole of Na and 2.82 moles of chlorine gas. The balanced equation indicates that 1 mole of Na will react with 1/2 mole of Cl_2 gas to produce 1 mole of NaCl. Therefore, there is an excess of Cl_2 (2.82 - 0.50 = 2.32 moles) that will go unreacted because there will not be enough Na to react with it – that is, Na is the limiting reagent in these reaction conditions. The extent of the reaction is limited by the Na that is available. At the end of the reaction, all the Na will be used up, one mole of NaCl will be produced and 2.32 moles of Cl_2 will remain unreacted.

SELF-TEST QUESTIONS

MULTIPLE CHOICE. In the following exercises, select the correct answer from the choices listed. In some cases, two or more answers will be correct.

1. The reaction of C_4H_{10} with O_2 to produce CO_2 and H_2O is what type of reaction?

    a. Precipitation
    b. Combustion

    c. Rusting
    d. Respiration

2. The formula weight for $Ca_3(PO_4)_2$ is:

    a. 279.3
    b. 310.3

    c. 167.3
    d. 543.3

3. A thimble of water contains 4.0×10^{21} molecules. The mass of water is:

    a. 1.1 g
    b. 11 g

    c. 1.1×10^{-8} g
    d. 0.11 g

4. Calculate the number of moles of aspirin, $C_9H_8O_4$, in a 4.0 g tablet.

    a. 2.3×10^{-4}
    b. 2.3

    c. 4.6×10^{-3}
    d. 0.023

5. The number of grams of K_2O produced from the oxidation of 4.0 grams of K is:
   (first write out a balance reaction)

    a. 4.7 g
    b. 4.0 g

    c. 9.4 g
    d. 6.3 g

6. If the reaction in problem 5 actually produces only 1.0 g K_2O, what is the percentage yield for the reaction?

    a. 25%
    b. 11%

    c. 21%
    d. 16%

7. In the equation below, the oxidizing agent in the reaction is:

$$Al_{(s)} + 2\,H^+ = Al^{+3} + H_{2(g)}$$

    a. $Al_{(s)}$
    b. H^+

    c. Al^{+3}
    d. H_2

8. The species that is oxidized in problem 7 is:

    a. $Al_{(s)}$ c. Al^{+3}
    b. H^+ d. H_2

9. Indicate which reaction is the most endothermic.

    a. $A + B \rightarrow C + 1,600$ cal.
    b. $D + E + 12,000$ cal. $\rightarrow F$
    c. $G + H \rightarrow I + 2.3$ kcals.
    d. $X + Y + 2.6$ kcal. $\rightarrow Z$

10. A chemist used a piece of laboratory equipment called a calorimeter to determine that the combustion of glucose produced 686 kcal per mole of glucose. The same reaction takes place in the body (a reaction in our metabolism). The expected amount of energy produced in the reaction in the body is:

    a. greater than 686 kcal.   c. equal to 686 kcal.
    b. less than 686 kcal.   d. can not predict

11. The enamel in your teeth contains about 95% hydroxyapatite, which has the formula, $Ca_{10}(PO_4)_6(OH)_2$. Fluoridation of water brings fluoride into your mouth and the fluoride ions exchange with the OH^- ion to form fluoroapatite, $Ca_{10}(PO_4)_6F_2$. This is widely used because fluoroapatite is:

    a. a tooth whitener   c. more insoluble in the acidic saliva
    b. a breath freshener   d. more soluble in the acidic saliva

COMPLETION. Write the word, phrase or number in the blank that will complete the statement or answer the question.

1. The mass of germanium, Ge, (atomic number 32 and atomic weight 72.59) which contains a mole of Ge atoms is _____.

2. A _____ is used to determine the _____ value of foods.

3. The _____ _____ _____ is measured in a calorimeter. This is determined by measuring the temperature change of the water in the calorimeter. If the temperature decreases as a result of the reaction, the reaction is _____.

4. An _____ reaction is one in which _____ is given off, but not necessarily as _____.

5. _____ in organic compounds is defined as the gain of hydrogen or the loss of _____.

55

6. Four very common reaction types in aqueous solution which involve ions are (1) _____, (2) _____, (3) _____ or (4) _____.

7. Complete the following table.

Compound	Grams	Number of Moles	Molecules
a. NH_3	_____	6.0×10^{-4}	_____
b. $C_2H_2Br_2$	6.3	_____	_____
c. $Si_2C_6H_{18}$	_____	_____	9.0×10^{12}

8. Complete and balance the following equations.

a. $C_5H_{12} + O_2 \rightarrow CO_2 + H_2O$

b. $Al^{+3} + SO_4^{2-} \rightarrow$

c. $SnO_2 + H_2 = Sn_{(s)} + H_2O$

d. $Au + Cl_2 \rightarrow AuCl_3$

e. $C_6H_{12}N_4 + H_2O \rightarrow H_2CO + NH_3(g)$

f. $HBr + KOH \rightarrow KBr + H_2O$

9. Indicate the type of reaction that is written out in question 8, a-d and f.

10. Consider the reaction of X and Y to form Z, with the indicated stoichiometry.

$$X + 2Y \rightarrow Z$$

The formula weights are: X = 50; Y = 200; Z = 100

a. The reaction indicates that _____ moles of X react with _____ moles of Y to produce _____ moles of Z.

b. If 200 grams of X is consumed in the reaction, the number of grams of Y consumed is _____ and the number of moles of Z produced is _____.

11. Consider the same stoichiometric equation in question 9. If 3 moles of Z were formed in the reaction, then _____ grams of X reacted with _____ grams of Y.

56

12. Consider the reaction of magnesium metal with oxygen gas.

$$Mg + O_2 = MgO$$

The reaction is carried out using 24.3 g of Mg and 32 g of O2. The limiting reagent, which will be completely used up, is _____ . The reaction will produce _____ g of MgO and leave an excess of _____ g of the unreacted _____.

QUICK QUIZ

Indicate whether the statements are true or false. The number of the question refers to the section of the book that the question was taken from.

4.2 What Are Formula Weights and Molecular Weights?
(a) Formula weight is the mass of a compound expressed in grams.
(b) One atomic mass unit (amu) is equal to one gram (g).
(c) The formula weight of H_2O is 18 amu.
(d) The molecular weight of H_2O is 18 amu.
(e) The molecular weight of a covalent compound is the same as its formula weight.

4.3 What Is a Mole and How Do We Use It to Calculate Mass Relationships?
(a) The mole is a counting unit, just as a dozen is a counting unit.
(b) Avogadro's number is the number of formula units in one mole.
(c) Avogadro's number, to three significant figures, is 6.02×10^{23} formula units per mole.
(d) One mole of H_2O contains $3 \times 6.02 \times 10^{23}$ formula units.
(e) One mole of H_2O has the same number of molecules as one mole of H_2O_2.
(f) The molar mass of a compound is its formula weight expressed in amu.
(g) The molar mass of H_2O is 18 g/mol.
(h) One mole of H_2O has the same molar mass as one mole of H_2O_2.
(i) One mole of ibuprofen, $C_{13}H_{18}O_2$, contains 33 moles of atoms.
(j) One mole of H_2O contains the same number of molecules as one mole of H_2O_2.
(k) To convert moles to grams, multiply by Avogadro's number.
(l) To convert grams to moles, multiple by molar mass.
(m) One mole of H_2O contains one mole of hydrogen atoms and one mol of oxygen atoms.
(n) One mole of H_2O contains 2 g of hydrogen atoms and 1 g of oxygen atoms.

4.4 How Do We Balance Chemical Equations?
(a) A balanced chemical equation shows the number of moles of starting materials reacting and the number of moles of products formed.
(b) In a chemical reaction, the number of moles of products always equals the number of moles of starting materials.
(c) In a chemical reaction, the mass of the products always equals the mass of the starting materials that have reacted.
(d) In a chemical reaction, the number of atoms in the products always equals the number of atoms in the starting material.
(e) In a balanced chemical equation, the grams of products always equals the grams of starting materials.

57

(f) The coefficients in a balanced chemical equation show the ratios of the moles of each product to the moles of each starting material.

4.5 How Do We Calculate Mass Relationships in Chemical Reactions?

(a) Stoichiometry is the study of mass relationships in a chemical reaction.

(b) To determine mass relationships in a chemical reaction, you first need to know the balanced chemical equation for the reaction.

(c) To convert from grams to moles, and vice versa, use Avogadro's number as a conversion factor.

(d) To convert from grams to moles, and vice versa, use molar mass as a conversion factor.

(e) A limiting reagent is the reagent that is used up in the reaction.

(f) If a chemical reaction between A and B requires $1A + 2B$, then A is the limiting reagent.

(g) Suppose a chemical reaction between A and B requires 1 mol of A and 2 mol of B. If 1 mol of each is present, then B is the limiting reagent.

(h) Theoretical yield is the yield of product that should be obtained according to the balanced chemical equation.

(i) Theoretical yield is the yield of product that should be obtained if all limiting reagent is converted to product.

(j) Percent yield is the number of grams of product divided by the number of grams of limiting reagent times 100.

(k) To calculate percent yield, divide the mass of product formed by the theoretical yield and multiply by 100.

4.6 How Can We Predict if Ions in Aqueous Solution Will React with Each Other?

(a) A net ionic equation shows only those ions that undergo chemical reaction.

(b) In a net ionic equation, the number of moles of starting material must equal the number of moles of product.

(c) A net ionic equation must be balanced in terms of both mass and charge.

(d) As a generalization, all lithium, sodium, and potassium salts are soluble in water.

(e) As a generalization, all nitrate (NO_3^-) salts are soluble in water.

(f) As a generalization, most carbonate (CO_3^{2-}) salts are insoluble in water.

(g) Sodium carbonate, Na_2CO_3, is insoluble in water.

(h) Ammonium carbonate, $(NH_4)_2CO_3$, is insoluble in water.

(i) Calcium carbonate, $CaCO_3$, is insoluble in water.

(j) Sodium dihydrogen phosphate, NaH_2PO_4, is insoluble in water.

(k) Sodium hydroxide, $NaOH$, is soluble in water.

(l) Barium hydroxide, $Ba(OH)_2$, is soluble in water.

4.7 What Is Oxidation and Reduction?

(a) When a substance is oxidized, it loses electrons.

(b) When a substance is reduced, it gains electrons.

(c) In a redox reaction, the oxidizing agent becomes reduced.

(d) In a redox reaction, the reducing reagent becomes oxidized.

(e) When Zn is converted to Zn^{2+} ion, we say that zinc is oxidized.

(f) Oxidation can be defined as the loss of oxygen atoms and/or the gain of hydrogen atoms.

(g) Reduction can be defined as the gain of oxygen atoms and/or the loss of oxygen atoms.

(h) When oxygen, O_2, is converted to hydrogen peroxide, H_2O_2, we say that O_2 is reduced.

(i) Hydrogen peroxide, H_2O_2, is an oxidizing agent.

(j) All combustion reactions are redox reactions.

(k) The products of complete combustion (oxidation) of hydrocarbon fuels are carbon dioxide and water.

(l) In the combustion of hydrocarbon fuels, oxygen is the oxidizing agent and the hydrocarbon fuel is the reducing agent.

(m) Incomplete combustion of hydrocarbon fuels can produce significant amounts of carbon monoxide.

(n) Most common bleaches are oxidizing agents.

4.8 What is Heat of Reaction?

(a) Heat of reaction is the heat given off or absorbed by a chemical reaction.

(b) An endothermic reaction gives off heat.

(c) If a chemical reaction is endothermic, the reverse reaction is exothermic.

(d) All combustion reactions are exothermic.

(e) If the reaction of glucose ($C_6H_{12}O_6$) and O_2 in the body to give CO_2 and H_2O is an exothermic reaction, then photosynthesis in green plants (the reaction of CO_2 and H_2O to give glucose and O_2) is an endothermic process.

(f) The energy required to drive photosynthesis comes from the Sun in the form of electromagnetic radiation.

ANSWERS FOR FOCUSED REVIEW EXERCISES

Exercise I.

a.	223.2		e.	48.0
b.	62.0		f.	284.0
c.	204.0		g.	89.0
d.	295.5		h.	300.0

Exercise II.

A.   (i)   <u>1 mole HF</u>
            20 g HF   ;   0.20 moles

    (ii)   <u>1 mole KOH</u>
            56.1 g KOH ;   0.59 moles

    (iii)   <u>1 mole AlF$_3$</u>
            84 g AlF$_3$; 2.4×10^{-4} moles

    (iv)   <u>137.4 g PCl$_3$</u>
            1 mole PCl$_3$;   41 g PCl$_3$

    (v)   <u>76.6 g GeH$_4$</u>
            1 mole GeH$_4$;   92 g GeH$_4$

(vi) $\dfrac{46\text{ g H}_2\text{CO}_2}{1\text{ mole H}_2\text{CO}_2}$;   0.46 g H_2CO_2

B.   (i) $\dfrac{6.02\times10^{23}\text{ molecules}}{1\text{ mole H}_2}$; 2.0×10^{23} molecules

(ii) $\dfrac{6.02\times10^{23}\text{ molecules}}{1\text{ mole Na}_2\text{SO}_4}$; 6.8×10^{24} molecules

(iii) $\dfrac{6.02\times10^{23}\text{ molecules}}{1\text{ mole NH}_3}$;   2.0 molecules

(iv) $\dfrac{1\text{ mole I}_2}{6.02\times10^{23}\text{ molecules}}$; 4.2×10^{-22} g.

(v) $\dfrac{1\text{ mole O}_3}{6.02\times10^{23}\text{ molecules}}$;   0.48 g.

(vi) $\dfrac{1\text{ mole HClO}_2}{6.02\times10^{23}\text{ molecules}}$; 4.6×10^{-16} g.

C.   (i) $\left(\dfrac{1\text{ mole}}{120.4\text{ g NCl}_3}\right)\left(\dfrac{6.02\times10^{23}\text{ molecules}}{1\text{ mole}}\right)$; 7×10^{20} molecules

(ii) $\left(\dfrac{1\text{ mole}}{256.6\text{ g S}_8}\right)\left(\dfrac{6.02\times10^{23}\text{ molecules}}{1\text{ mole}}\right)$; 9×10^{18} molecules

(iii) $\left(\dfrac{1\text{ mole}}{127.9\text{ g HI}}\right)\left(\dfrac{6.02\times10^{23}\text{ molecules}}{1\text{ mole}}\right)$; 3×10^{16} molecules

(iv) $\left(\dfrac{149.9\text{ g SeCl}_2}{1\text{ mole}}\right)\left(\dfrac{1\text{ mole}}{6.02\times10^{23}\text{ molecules}}\right)$; 9.9×10^5 g

(v) $\left(\dfrac{60\text{ g C}_2\text{H}_4\text{O}_2}{1\text{ mole}}\right)\left(\dfrac{1\text{ mole}}{6.02\times10^{23}\text{ molecules}}\right)$; 6×10^{-10} g

(vi) $\left(\dfrac{580.02\text{ g GeI}_4}{1\text{ mole}}\right)\left(\dfrac{1\text{ mole}}{6.02\times10^{23}\text{ molecules}}\right)$; 1.9×10^{-19} g

Exercise III.

A.   1.1 moles O_2; 0.7 moles CO_2; 0.8 moles H_2O

B. 9×10^{-3} moles C_7H_{16}; 6×10^{-2} moles CO_2; 7×10^{-2} moles H_2O

C.   0.01 moles C_7H_{16}; 0.16 moles O_2; 0.11 moles H_2O

ANSWERS TO SELF-TEST QUESTIONS

MULTIPLE CHOICE

1.	b	5.	a	9.	b
2.	b	6.	c	10.	c
3.	d	7.	b	11.	c
4.	d	8.	a		

COMPLETION

1. 72.59 g
2. calorimeter, caloric
3. heat of reaction, endothermic
4. exergonic, energy, heat
5. reduction, oxygen
6. precipitation, gas formation, acid neutralization of a base, redox

7.   a.   0.010 g, 3.6×10^{20} molecules

     b. 3.4×10^{-2} moles, 2.0×10^{22} molecules

     c.   2.2 ng, 1.5×10^{-11} moles

8.   a. $C_5H_{12} + 8O_2 \rightarrow 5CO_2 + 6H_2O$

     b. $2Al^{+3} + 3SO_4^{-2} \rightarrow Al_2(SO_4)_3$

     c. $SnO_2 + 2H_2 = Sn_{(s)} + 2H_2O$

     d. $Au + 3/2Cl_2 = AuCl_3$

     e. $C_6H_{12}N_4 + 6H_2O = 6H_2CO + 4NH_3$

     f. $HBr + KOH \rightarrow KBr + H2O$ (balanced as written)

9.   a.   combustion
     b.   combination
     c.   single displacement
     d.   oxidation and reduction (redox) reaction
     e.   gas production
     f.   neutralization of a base (KOH) with an acid (HBr)

61

10. a. one, two, one
     b. 1600 g, 4 moles

11. 150 g, 1200 g

12. Mg, 40.3 g, 16 g, O_2

QUICK QUIZ

4.2

(a) False	(b) False	(c) True	(d) True
(e) True			

4.3

(a) True	(b) True	(c) True	(d) False
(e) True	(f) False	(g) True	(h) False
(i) True	(j) True	(k) False	(l) False
(m) False	(n) False		

4.4

(a) True	(b) False	(c) True	(d) True
(e) True	(f) True		

4.5

(a) True	(b) True	(c) False	(d) True
(e) True	(f) False	(g) True	(h) True
(i) True	(j) False	(k) True	

4.6

(a) True	(b) False	(c) True	(d) True
(e) True	(f) True	(g) False	(h) False
(i) True	(j) False	(k) True	(l) True

4.7

(a) True	(b) True	(c) True	(d) True
(e) True	(f) False	(g) False	(h) True
(i) True	(j) True	(k) True	(I) True
(m) True	(n) True		

4.8

(a) True	(b) False	(c) True	(d) True
(e) True	(f) True		

62

Happy is he who gets to know the
reasons for things.
- Motto of Churchill College,
Cambridge

5

Gases, Liquids and Solids

CHAPTER OBJECTIVES

After studying the chapter and working the assigned exercises in the text and study guide, you should be able to do the following.

1. State the basic assumptions in the kinetic-molecular theory of gases. Indicate how each individual assumption helps in understanding the behavior or properties of ideal gases.

2. Explain how a barometer and a manometer are setup to read gas pressure.

3. Name the four properties of a gas which defines its physical state and indicate the common units for each.

4. Apply Boyle's Law ($P_1V_1 = P_2V_2$, conditions of constant temperature) to calculate the final volume or pressure of a gas.

5. Use Charles's Law ($V_1/T_1 = V_2/T_2$, conditions of constant pressure) to calculate the final volume or temperature of a gas.

6. Utilize Gay-Lussac's Law ($P_1/T_1 = P_2/T_2$, conditions of constant volume) to calculate the final temperature or pressure of a gas.

7. Apply the combined gas law ($P_1V_1/T_1 = P_2V_2/T_2$) to calculate the final pressure, temperature or volume of a gas.

8. Use the ideal gas law (PV = nRT) to calculate either the pressure, volume, the number of moles or the temperature of the gas.

9. Utilize the modified form of the ideal gas law to calculate the molecular weight or the mass of a quantity of gas.

10. Utilize Avogadro's Law (PV/RT = n, at conditions of a specific pressure and temperature) to show that equal volumes of gases at the same temperature and pressure contain equal numbers of gas molecules.

11. Apply Dalton's Law of Partial Pressures, ($P_T = P_1+P_2+P_3...$), to calculate the partial pressure of a gas in a mixture or the total pressure in a mixture of gases.

12. State the six basic tenets of the kinetic molecular theory of gases.

13. Name, define and give the approximate energies involved in the three basic intermolecular forces between molecules.

14. Draw a few molecules that can participate in intermolecular hydrogen bonding and show the specific interactions involved.

15. Define boiling point and explain the factors that influence the temperature at which a liquid boils.

16. Indicate the specific make-up and the general characteristics of an a) ionic, b) molecular, c) polymeric, d) network and e) amorphous solid.

17. Draw a heating curve for ice, explain the main features of the curve and explain the significance of the heat of fusion and the heat of vaporization in this figure.

18. Draw the phase diagram for water, pointing out the triple point and indicating the regions where only solid, liquid or gas exists.

19. Define the important terms and comparisons in this chapter and give specific examples where appropriate.

Make a summary sheet of the important laws and corresponding equations presented in this chapter that provide the basis for characterizing the property of a gas. Indicate the parameters that are kept constant in each relationship and the units of each.

IMPORTANT TERMS AND COMPARISONS

Gaseous, Liquid and Solid State

Melting Point and Freezing Point

Barometer and Manometer

Arterial and Venous Blood

London Dispersion Force, Hydrogen Bonds
and Dipole-Dipole Interactions

Intermolecular Forces
Entropy
Boyle's, Charles's and Gay-Lussac's Law
Avogadro's Law
Standard Temperature and Pressure (STP)
Vapor Pressure of a Liquid
Systolic and Diastolic Pressure
Sphygmomanometer
Crystals, Amorphous and Network Solids
Heating Curve
Heat of Fusion and Heat of Vaporization
Sublimation
Nanotubes

Torr, mm Hg & Pascals
Atmosphere (Unit of Pressure)
Dalton's Law of Partial Pressures
Ideal Gas Law
Equilibrium
Partial and Total Pressure
Boiling Point and Melting Point
Crystallization
Phases and Phase Changes
Specific Heat
Phase Diagram and Triple Point
Buckminsterfullerene or "Bucky Ball"

FOCUSED REVIEW

One of the most perplexing difficulties that many students encounter in problem solving is deciding the equation to use to obtain the answer. Since there are a number of equations (relationships) that may be used in solving gas law problems, a simple strategy is necessary to find just the right relationship for the particular problem.

The parameters that may be involved in a particular problem are the following:

Parameter	Common Units
Pressure (P)	Atmosphere, mm Hg, torr, pascal
Volume (V)	Liter, milliliter
Temperature (T)^{o}C, K	
Moles (n)	--------
Mass (m)	Grams
Molecular Weight (M.W.)	Atomic Mass Units

The types of problems can be subdivided into two major categories. The first is the situation in which the gas is completely characterized in one state, the **conditions are CHANGED**, and one of the characteristics of the second state is unknown. A number of situations may arise.

TABLE I

	State 1		State 2	
	Known Quantity		Known Quantity	Unknown Quantity

Ideal Gas Law:

$$\frac{P_1V_1}{T_1} = \frac{P_2V_2}{T_2}$$

$(P_1, V_1 T_1)$	1. P_2V_2	T_2	or
	2. P_2, T_2	V_2	or
	3. V_2, T_2	P_2	

Boyle's Law:

$$P_1V_1 = P_2V_2, \text{ at constant } T$$

| (P_1, V_1) | 1. P_2 | V_2 | or |
| | 2. V_2 | P_2 | |

Charles's Law:

$$\frac{V_1}{T_1} = \frac{V_2}{T_2}, \text{ at constant } P$$

| (V_1, T_1) | 1. V_2 | T_2 | or |
| | 2. T_2 | V_2 | |

Gay-Lussac's Law:

$$\frac{P_1}{T_1} = \frac{P_2}{T_2}, \text{ at constant } V$$

| (P_1, T_1) | 1. P_2 | T_2 | or |
| | 2. T_2 | P_2 | |

The second category of problems involves a gas or gases in one **PARTICULAR** state. **No changes in conditions are involved.** In this case, a specific property in this state is to be calculated, using either the ideal gas law, Dalton's Law of partial pressures, Avogadro's Law or Graham's Law of diffusion.

Recall that the ideal gas law equation may be written in two forms.

(1) $PV = nRT$

and since

$$n = \frac{g}{M.W.}$$

(2) $PV = \dfrac{gRT}{(M.W.)}$

66

Therefore, one can solve for any one of the parameters in equation (1) or (2), if the other quantities are given in the problem.

Two final points of interest.

1.   R, the universal gas constant, is a physical quantity, which you'll recall has both a numerical value and **units**. The value of R presented in the text is R = 0.082 L-atm/mole-K. Therefore, it is **essential** when using this equation with this value of R that the gas parameters conform with these units of R. That is, the parameters must have units of P (atm), V (L), n (moles) and T (K). If the parameters are not given in the proper units, convert them immediately. Be on the outlook for just this situation.

2.   In problems involving two states of a gas and using any one of the four gas laws presented, it is also essential that the units characterizing state 1 be the same as in state 2. For example, if the pressure in state 1 is given as 1.2 atm, while the pressure in state 2 is given as 500 mm Hg, one of these values (usually the 500 mm Hg value) must be converted to the other before the calculation.

Knowing these preliminaries and precautions, the procedure for analyzing and solving a gas law problem is:

1. Determine if the problem involves the gas in only one state or two different states.

2. Determine (**write out**) what is known and what is to be determined (unknown) in the problem.

3. From your analysis in steps 1 and 2, decide which law and equation is appropriate to solve the problem.

4. Rearrange the equation if necessary and check that all units are appropriate.

5. Solve for the unknown quantity.

To pull some of this together, consider the following problem.

If a sample of 1.78 moles of $CHCl_3$ gas occupies 68 L at a pressure of 275 mm Hg, what is its temperature in °C?

Step 1.   The problem involves $CHCl_3$ gas in **one state**. There are no changes to a new state. Therefore the equations as written in Table I (Study Guide) are not appropriate and either the (1) ideal gas law, (2) Dalton's Law of partial pressures or (3) Avogadro' Law must be used.

Step 2.   Knowns   Unknown

               n = 1.78 moles
               V = 68 L   T = ?

P = 275 mm Hg

Step 3. From the parameters given and the unknown quantity, the ideal gas law should be used in the form.

$$PV = nRT$$

Step 4. Rearrange by dividing through each side by nR. This yields

$$T = \frac{PV}{nR}$$

Convert pressure (P = 275 mm Hg) into units of atmosphere to conform with units of R, the universal gas constant.

$$P = \left(\frac{275 \text{ mm Hg}}{1}\right)\left(\frac{1 \text{ atm}}{760 \text{ mm Hg}}\right) = 0.362 \text{ atm}$$

Step 5. $$T = \frac{(0.362 \text{ atm})(68 \text{ L})}{(1.78 \text{ moles})(0.082 \text{ L-atm/mole-K})} = 169 \text{ K}$$

The final step in this problem is to convert the temperature from K to °C.
°C = 273° + K = 273° + 169,

Therefore, the final answer is: **T = 442°C**

SELF-TEST QUESTIONS

<u>MULTIPLE CHOICE</u>. Select the correct answer from the choices given. Only one of the choices is correct.

1. A closed aerosol-can that had been filled with shaving cream contains 3.0 g of nitrogen gas (N_2) as the propellant. If the volume of the can as 200 mL, the pressure in the can at 25°C is:

   a.   1.1 atm   c.   1.3 atm
   b.   130 atm   d.   13 atm

2. If the can in question 1 is thrown into an incinerator or fire, the temperature in the can could well get as high as 850 K. The pressure in the can, assuming that it did not explode, would be:

   a.   37 atm   c.   370 atm
   b.   3.7 atm   d.   370 mm Hg

68

3. A gas, which is initially contained in a movable cylinder of 900 mL volume and at 25°C, exerts a pressure of 3 atmospheres. The pressure is reduced at constant temperature to a value of 380 mm Hg. The new volume is:

   a. 5400 L
   b. 5.4 L

   c. 7.1×10^{-3} L
   d. 7.1 L

4. A piece of Dry Ice (solid carbon dioxide, CO_2) is placed in a 1 L tank. It sublimes to CO_2 gas at STP. The amount of CO_2 was:

   a. 1 mg
   b. 44 g

   c. 4.4 g
   d. 2 g

5. The average kinetic energy of molecules is proportional to the

   a. temperature in °C
   b. velocity of the molecule

   c. temperature in K
   d. volume at STP

6. The official S1 unit of pressure is:

   a. torr
   b. pascal

   c. atmosphere
   d. mm Hg

7. The autoclave operates on the principle of:

   a. Dalton's Law of Partial Pressure
   b. Boyle's Law
   c. Gay-Lussac's Law
   d. Charles's Law

8. A gas at 25°C is initially contained in a movable cylinder of 2000 mL volume and exerts a pressure of 38 mm Hg. The pressure is changed to 0.5 atm and the volume of the cylinder increases to 2.2 L. The final temperature in the cylinder is:

   a. 3×10^3 K
   b. 275 K

   c. 3.2 K
   d. 2×10^4 K

9. Consider the reaction: $3K + Y \rightarrow 2Z$. At STP, 2 L of Z are produced from the reaction of X and Y. The number of liters of X necessary is:

   a. 1 L
   b. 2 L

   c. 3 L
   d. cannot determine

69

10. A quantity of 6 grams of a gas occupies 300 milliliters at STP. The molecular weight of the gas is:

    a. 440   c. 44
    b. 660   d. 66

11. Predict which liquid would exhibit the greatest surface tension.

    a. CCl3H   c. CH3OH
    b. H2O   d. CCl4

12. Indicate which of the following are characteristics of "buckyballs". There may be more than one answer.

    a. Contains 60 carbon atoms
    b. Carbons are arranged in pentagons and octagons
    c. Contains only carbon atoms
    d. Each carbon is bonded to four other carbons

13. Indicate which of the following are isoforms of carbon. There may be more than one answer.

    a. diamond   c. graphite
    b. soot   d. fullerenes

14. You have 100 g of ice and 100 g of water. What can be said about the volume of the ice and the water?

    a. both are the same   c. water has a greater volume
    b. ice has a greater volume   d. cannot determine or predict

COMPLETION. Write the word, phrase or number in the blank that will complete the statement or answer the question.

1. STP indicates conditions in which T = _____ and P = _____.

2. The boiling point of HF (M.W. = 18) is higher than that for CH_4 (M.W. = 18) because of _____, while the boiling point of octane (C_8H_{18}) is higher than that for propane (C_3H_8) because of the _____.

3. The behavior of an ideal gas at constant volume but at varying pressure or temperature is governed by _____ Law.

4. The O_2 gas in a 1 L tank at 1 atmosphere is transferred to a 2 L tank that already contains 2 atmospheres of N_2 gas. The final pressure is _____.

5. The higher the temperature, the _____ the _____ energy of molecules. Therefore, by heating a closed container of gas, the pressure will _____.

6. At STP, a 5.6 L container of Ne or O_2 contains _____ moles of gas or _____ molecules.

7. The energy required to break a hydrogen bond is approximately _____ kcal/mole.

8. The atmospheric pressure in Denver is less than in Bowling Green, Ohio. The boiling point of water in Bowling Green will be _____ than the boiling point of water in Denver. This is understandable from the definition for the boiling point of a liquid which states that the boiling point of a liquid is the temperature at which the _____ _____ of the liquid is equal to the _____ _____.

9. The basic postulates in the kinetic molecular theory of gases are:

   i. _____

   ii. _____

   iii. _____

   iv. _____

   v. _____

   vi. _____

10. The melting point of diamond, which is a _____ _____, is much higher than that for a _____ crystal.

11. The two plateaus in a heating curve represent the temperatures at which two phases _____. The amount of heat necessary to convert one gram of water at 100°C to steam is called the _____.

12. The intermolecular forces responsible for the solidification of Xe and other noble gases are _____.

13. The unit of pressure, _____, is equivalent to the _____.

14. Convert the physical quantities in the left column to their equivalent in the right-hand column.

    a. 4.3 mL   _____ L

    b. 34ºC   _____ K

    c. 450 mm Hg   _____ atm

d.   43 g O_2 _____ moles O_2

e.   372 K _____ °C

f.   STP _____ mm Hg

_____ °C

g.   17 cal/mol _____ Kcal/mol

15.   Match up the unit or instrument in the column at the right with its use in the left-hand column.

Unit or Instrument   Use

a.   Barometer   i.   Measure blood pressure
b.   Sphygmomanometer   ii.   Laboratory instrument used to measure pressure
c.   Closed container of liquid   iii. Hospital unit used for patients with CO
d.   Hyperbaric Chamber   poisoning, smoke inhalation, in addition to
e.   Manometer   other uses
                                   iv.   Measures atmospheric pressure
                                   v.   Used in determination of vapor pressure

16.   Water has a superstructure involving intermolecular _____, which in part explains why the density of _____ is greater than that of _____.

17.   As the temperature decreases, the vapor pressure of a liquid _____.

18.   _____ is the transition from the solid state directly into the _____ state without going through the _____ state.

19.   An amorphous solid differs from a crystalline solid in that the molecules or ions are solidified in a _____ pattern.

20.   Write the equation for the behavior of an ideal gas. _____.

21.   A new form of carbon, which contains _____ carbons and has a structure which resembles a soccer ball is called either _____ or _____.

22.   A measure of order is called _____. As the temperature of a substance increases in temperature, its _____ value will _____.

23.   Examples of solids that can be described as molecular, polymeric or network are _____, _____, and _____, respectively.

72

24. Two types of solids that exhibit very high melting points are _____ and _____. Sodium chloride (NaCl) is an example of a _____ solid.

25. A phase diagram plots _____ on the x axis and _____ on the y axis.

26. The line in a phase diagram that separates the solid phase from the liquid phase contains all the _____ points for that substance.

27. The _____ _____ is a unique condition in which all three phases coexist in a phase diagram.

QUICK QUIZ
    Indicate whether the statements are true or false. The number of the question refers to the section of the book that the question was taken from.

5.2 What Is Gas Pressure and How Do We Measure It?
    (a) Gas pressure can be measured by both a barometer and a manometer.
    (b) One atmosphere is equal to 760 mm Hg.
    (c) At sea level, the average pressure of the atmosphere is 28.9 in. Hg.

5.3 What Are the Laws That Govern the Behavior of Gases?
    (a) For a sample of gas at constant temperature, its pressure multiplied by its volume is a constant.
    (b) For a sample of gas at constant temperature, increasing the pressure increases the volume.
    (c) For a sample of gas at constant temperature, $P_1/V_1 = P_2/V_2$.
    (d) As a gas expands at constant temperature, its volume increases.
    (e) The volume of a sample of gas at constant pressure is directly proportional to its temperature—the higher its temperature, the larger its volume.
    (f) A hot-air balloon rises because hot air is less dense than cooler air.
    (g) For a gas sample in a container of fixed volume, an increase in temperature results in an increase in pressure.
    (h) For a gas sample in a container of fixed volume, P x T is a constant.
    (i) When steam at 100°C in an autoclave is heated to 120°C, the pressure within the autoclave increases.
    (j) When a gas sample in a flexible container at constant pressure at 25°C is heated to 50°C, its volume doubles.
    (k) Lowering the diaphragm causes the chest cavity to increase in volume and the pressure of air in the lungs to decrease.
    (l) Raising the diaphragm decreases the volume of the chest cavity and forces air out of the lungs.

5.4 What Are Avogadro's Law and the Ideal Gas Law?
    (a) Avogadro's law states that equal volumes of gas at the same temperature and pressure contain equal numbers of molecules.
    (b) At STP, one mole of uranium hexafluoride (UF_6, MW 352 amu), the gas used in uranium enrichment programs, occupies a volume of 352 L.

73

(c) If two gas samples have the same temperature, volume, and pressure, then both contain the same number of molecules.

(d) The value of Avogadro's number is 6.02×10^{23} grams.

(e) Avogadro's number is valid only for gases at STP.

(f) The ideal gas law is $PV = nRT$.

(g) When using the ideal gas law for calculations, temperature must be in degrees Celsius.

(h) If one mole of ethane (CH_3CH_3) gas occupies 20.0 L at 1.00 atm, the temperature of the gas is 244 K.

(i) One mole of helium (MW 4.0 amu) gas at STP occupies twice the volume of one mole of hydrogen (MW 2.0 amu).

5.5 What Is Dalton's Law of Partial Pressures?

(a) Partial pressure is the pressure that a gas in a container would exert if it were alone in the container.

(b) The units of partial pressure are grams per liter.

(c) Dalton's law of partial pressures states that the total pressure of a mixture of gases is the sum of the partial pressures of each gas.

(d) If 1 mole of CH_4 gas at STP is added to 22.4 L of N_2 at STP, the final pressure in the 22.4-L container will be 1 atm.

5.6 What is the Kinetic Molecular Theory?

(a) According to the kinetic molecular theory, gas particles have mass but no volume.

(b) According to the kinetic molecular theory, the average kinetic energy of gas particles is proportional to the temperature in degrees Celsius.

(c) According to the kinetic molecular theory, when gas particles collide, they bounce off each other with no change in total kinetic energy.

(d) According to the kinetic molecular theory, there are only weak intramolecular forces of attraction between gas particles.

(e) According to the kinetic molecular theory, the pressure of a gas in a container is the result of collisions of gas particles with the walls of the container.

(f) Warming a gas results in an increase in the average kinetic energy of its particles.

(g) When a gas is compressed, the increase in its pressure is the result of an increase in the number of collisions of its particles with the walls of the container.

(h) The kinetic molecular theory describes the behavior of ideal gases, of which there are only a few.

(i) As the temperature and volume of a gas increase, the behavior of the gas becomes more like the behavior predicted by the ideal gas law.

(j) If the assumptions of the kinetic molecular theory of gases are correct, then there is no combination of temperature and pressure at which a gas would become liquid.

5.7 What Types of Attractive Forces Exist Between Molecules?

(a) Of the forces of attraction between particles, London dispersion forces are the weakest and covalent bonds are the strongest.

(b) All covalent bonds have approximately the same energy.

(c) London dispersion forces arise because of the attraction of temporary induced dipoles.

(d) In general, London dispersion forces increase as molecular size increases.

(e) London dispersion forces occur only between polar molecules—they do not occur between nonpolar atoms or molecules.

74

(f) The existence of London dispersion forces accounts for the fact that even small, nonpolar particles such as Ne, He, and H_2 can be liquefied if the temperature is low enough and the pressure is high enough.

(g) For nonpolar gases at STP, the average kinetic energy of its particles is greater than the force of attraction between gas particles.

(h) Dipole-dipole interaction is the attraction between the positive end of one dipole and the negative end of another dipole.

(i) Dipole-dipole interactions exist between CO molecules but not between CO_2 molecules.

(j) If two polar molecules have the same molecular weight, the strength of the dipole-dipole interactions between the molecules of each will be approximately the same.

(k) Hydrogen bonding refers to the single covalent bond between the two hydrogen atoms in H-H.

(l) The strength of hydrogen bonding in liquid water is approximately the same as that of an O-H covalent bond in water.

(m) Hydrogen bonding, dipole-dipole interactions, and London dispersion forces have in common that the forces of attraction between particles are all electrostatic (positive for negative and negative for positive).

(n) Water (H_2O, bp 100°C) has a higher boiling point than hydrogen sulfide (H_2S, bp -61°C) because the hydrogen bonding between H_2O molecules is stronger than that between H_2S molecules.

(o) The hydrogen bonding between molecules containing N-H groups is stronger than that between hydrogen molecules containing O-H groups.

5.8 How Do We Describe the Behavior of Liquids at the Molecular Level?

(a) The ideal gas law assumes that there are no attractive forces between molecules. If this were true, then there would be no liquids.

(b) Unlike a gas, whose molecules move freely in any direction, molecules in a liquid are locked into fixed positions, giving the liquid a constant shape.

(c) Surface tension is the force that prevents a liquid from being stretched.

(d) Surface tension creates an elastic-like layer on the surface of a liquid.

(e) Water has a high surface tension because H_2O is a small molecule.

(f) Vapor pressure is proportional to temperature— as the temperature of a liquid sample increases, its vapor pressure also increases.

(g) When molecules evaporate from a liquid, the temperature of the liquid drops.

(h) Evaporation is a cooling process because it leaves fewer molecules with high kinetic energy in the liquid state.

(i) The boiling point of a liquid is the temperature at which its vapor pressure equals the atmospheric pressure.

(j) As the atmospheric pressure increases, the boiling point of a liquid increases.

(k) The temperature of boiling water is related to how vigorously it is boiling—the more vigorous the boiling, the higher the temperature of the water.

(l) The most important factor determining the relative boiling points of liquids is molecular weight—the greater the molecular weight, the higher the boiling point.

(m) Ethanol (CH_3CH_2OH, bp 78.5°C) has a greater vapor pressure at 25°C than water (H_2O, bp 100°C).

(n) Hexane ($CH_3CH_2CH_2CH_2CH_2CH_3$, bp 69°C) has a higher boiling point than methane (CH_4, bp -164°C) because hexane has more sites for hydrogen bonding between its molecules than does methane.

(n) Hexane ($CH_3CH_2CH_2CH_2CH_2CH_3$, bp 69°C) has a higher boiling point than methane (CH_4, bp -164°C) because hexane has more sites for hydrogen bonding between its molecules than does methane.

(o) A water molecule can participate in hydrogen bonding through each of its hydrogen atoms and through its oxygen atom.

(p) For nonpolar molecules of comparable molecular weight, the more compact the shape of the molecule, the higher its boiling point.

5.9 What Are the Characteristics of the Various Types of Solids?
(a) Formation of a liquid from a solid is called melting; formation of a solid from a liquid is called crystallization.

(b) Most solids have a higher density than their liquid forms.

(c) Molecules in a solid are locked into fixed positions.

(d) Each compound has one and only one solid (crystalline) form.

(e) Diamond and graphite are both crystalline forms of carbon.

(f) Diamond consists of hexagonal crystals of carbon arranged in a repeating pattern.

(g) The "nano" in nanotube refers to the structure dimensions, which are in the nanometer (10^{-9} m) range.

(h) Nanotubes have lengths up to 1 nm.

(i) A buckyball (C_{60}) has a diameter of 1 nm.

(j) All solids, if heated to a high enough temperature, can be melted.

(k) Glass is an amorphous solid.

5.10 What Is a Phase Change and What Energies Are Involved?
(a) A phase change from solid to liquid is called melting.

(b) A phase change from liquid to gas is called boiling.

(c) If heat is added slowly to a mixture of ice and liquid water, the temperature of the sample gradually increases until all of the ice is melted.

(d) Heat of fusion is the heat required to melt 1 g of a solid.

(e) Heat of vaporization is the heat required to evaporate 1 g of liquid at the normal boiling point of the liquid.

(f) Steam burns are more damaging to the skin than hot-water burns because the specific heat of steam is so much higher than the specific heat of hot water.

(g) The heat of vaporization of water is approximately the same as its heat of fusion.

(h) The specific heat of water is the heat required to raise the temperature of 1 g of water from 0° to 100°C.

(i) Melting a solid is an exothermic process; crystallization of a liquid is an endothermic process.

(j) Melting a solid is a reversible process; the solid can be converted to a liquid and the liquid back to a solid with no change in composition of the sample.

(k) Sublimation is a phase change from solid directly to gas.

MULTIPLE CHOICE

1.	d	8.	a
2.	a	9.	c
3.	b	10.	a
4.	d	11.	b
5.	c	12.	a, c
6.	b	13.	a, b, c, d
7.	c	14.	b

COMPLETION

1. $T = 0°C$, $P = 1$ atm
2. intermolecular hydrogen bonds, (higher) molecular weight
3. Gay-Lussac's Law
4. 2.5 atm
5. greater, kinetic, increase
6. 0.25, 1.5×10^{23}
7. 5-10
8. greater, vapor pressure, atmospheric pressure
9. These statements can be paraphrased.
   i. A gas consists of molecules traveling in random directions, in straight lines and at a range of speeds.
   ii. The average kinetic energy, and therefore the average velocity of gas molecules, is directly proportional to the absolute temperature.
   iii. Gas molecules collide with each other and may exchange energy between each other, but the total kinetic energy is conserved.
   iv. The molecules in a gas take up essentially no volume. The volume of the gas is the volume of the container it is in.
   v. There are no attractions between the gas molecules.
   vi. The pressure of the gas is due to collisions of the molecules with the walls of the container. The greater the number of collisions per unit time, the greater the pressure.
10. network solid, molecular
11. coexist, heat of vaporization
12. London dispersion forces
13. torr, mm Hg or also 1 atm = 101,325 pascals is acceptable. The former equivalence is more commonly used, however.
14. a. 4.3×10^{-3} L
    b. 307 K
    c. 0.592 atm
    d. 1.3 moles O_2
    e. -1°C
    f. 760 torr, 0°C
    g. 1.7×10^{-2} kcal/mole

15. a. iv
    b. i
    c. v
    d. iii
    e. ii
16. hydrogen bonding, liquid water, ice
17. decreases
18. sublimation, gaseous, liquid
19. random
20. PV = nRT
21. 60; buckminsterfullerene; buckyballs
22. entropy, entropy, increase
23. sugar or ice; rubber, plastic or proteins; diamond or quartz
24. ionic, network, ionic
25. T, P
26. melting
27. triple point

QUICK QUIZ

5.2

(a) True	(b) True	(c) True

5.3

(a) True	(b) False	(c) False	(d) True
(e) True	(f) True	(g) True	(h) False
(i) True	(j) False	(k) True	(l) True

5.4

(a) True	(b) False	(c) True	(d) False
(e) False	(f) True	(g) False	(h) True
(i) False			

5.5

(a) True	(b) False	(c) True	(d) False

5.6

(a) True	(b) False	(c) True	(d) False
(e) True	(f) True	(g) True	(h) False
(i) True	(j) True		

5.7

(a) False	(b) False	(c) True	(d) True
(e) False	(f) True	(g) True	(h) True
(i) True	(j) False	(k) False	(l) False
(m) True	(n) True	(o) False	

5.8

(a) True	(b) False	(c) False	(d) True
(e) False	(f) True	(g) True	(h) True
(i) True	(j) True	(k) False	(l) False
(m) True	(n) False	(o) True	(p) False

5.9

(a) True	(b) True	(c) True	(d) False
(e) True	(f) False	(g) True	(h) False
(i) False	(j) False	(k) True	

5.10

(a) True	(b) True	(c) False	(d) True
(e) True	(f) False	(g) False	(h) False
(i) False	(j) True	(k) True	

The fluids of the human body and the seas of the
earth contain common table salt, sodium chloride,
and an assortment of other chemicals.
- S. Brooks
The Seas Inside Us: Water in the Life Processes

6

Solutions and Colloids

CHAPTER OBJECTIVES

After studying the chapter and working the assigned exercises in the text and study guide,
you should be able to do the following.

1. Summarize the characteristics of a mixture, a suspension, a colloidal dispersion and a true
   solution and give specific examples of each.

2. Indicate the factors which influence the solubility of a (1) gas, (2) liquid or (3) solid in a
   liquid solvent.

3. Perform calculations to determine the concentration of a solution expressed in (1) molarity,
   (2) % w/v, (3) % w/w and 4) % v/v. Given the concentration of a solution in any of these
   forms, convert to its equivalent in the other concentration terms.

4. Using the formula, moles = M_1 x V_1 = M_2 x V_2, calculate the concentration of solutions
   made by the dilution of a more concentrated solution.

5. Outline the characteristics of water that are most responsible for the unusual properties,
   such as (1) its high boiling point, (2) surface tension and (2) its ability to dissolve ionic
   compounds.

6. Specify the fundamental difference between a strong and weak electrolyte and a non-
   electrolyte.

7. Discuss the mechanisms by which water can effectively dissolve particular covalent
   molecules.

8. Considering both molecular and ionic compounds, carry out calculations to determine the freezing point depression produced by the various aqueous solutions.

9. Define osmosis and osmolarity and discuss how these concepts relate to specific areas of medicine.

10. List the characteristic differences between a solution, a colloid and a suspension.

11. Define the important terms and comparisons in this chapter and give specific examples where appropriate.

IMPORTANT TERMS AND COMPARISONS

Homogeneous and Heterogeneous Mixtures
Solute, Solvent and Solution
Clear vs. Colorless
Solvation and Hydration
A Physical Constant
Strong and Weak Electrolyte
Anhydrous and Hygroscopic
Saturated, Unsaturated and Supersaturated
    Solutions
"Like Dissolves Like"
Brownian Motion
Physiological Saline
Colligative Properties
Hemolysis and Crenation
(% w/v), (% w/w) and (% v/v)
Osmotic Pressure and Semipermeable Membrane
Colloidal Dispersions and Tyndall Effect
Parts per Million (ppm) and Part per Billion (ppb)

Alloy
Miscible and Immisible Liquids
Hydrates, Water of Hydrations and
    Plaster of Paris
Electrolyte and Non-Electrolyte
Cathode and Anode
Henry's Law
Concentration Terms
Solution and Suspension
Transparent, Translucent and Opaque
Solvation Layer
Dilute vs. Concentrated
Isotonic, Hypertonic and Hypotonic
Dialysis and Hemodialysis
Reverse Dialysis and Desalinization
Molarity and Osmolarity
Melting Point Depression

FOCUSED REVIEW OF CONCEPTS

In all chemically related fields, we are continuously called upon to make observations or measurements and to characterize a mixture or solution, either in a qualitative or quantitative sense. Recall that chemistry is an experimental science and our current knowledge is derived from **careful observations** of experimental happenings. Concepts in this chapter lay the foundation for such characterization. Consider, for example, that you are presented with a test tube containing some liquid and asked to characterize it as completely as you can. Your preliminary, yet very important, judgments will necessarily be **qualitative**. Initially, you might focus on these questions.

1. Does the test tube contain a true solution, a colloidal dispersion or a suspension?

2. If the test tube contains a true solution,
   a. Is the solution clear (transparent), translucent or opaque?
   b. Is the solution colored or colorless?
   c. Does the solution conduct electricity?
   d. Is the liquid simply a solvent or an actual solution?
   e. If it is a solution, is it dilute or concentrated and is it unsaturated, saturated or supersaturated with solute?

At this point, it should be evident that although the characteristics are qualitative, we know a great deal about the unknown solution in the test tube.

To characterize the solution further, however, requires a **quantitative analysis**. The following table collects a number of means of quantitatively analyzing solutions, with the definitions and examples of each. However, first just focus your attention on a basic point that is often times not clearly defined and leads to difficulty. You will note that all the quantitative concentration terms in Table I are expressed as the quantity of one reagent in a certain volume of **SOLUTION** (= solute plus solvent). The emphasis here is to distinguish the difference between the volume of the **total solution** from the volume of only the **solvent** used to dissolve the **solute**.

TABLE I

Quantitative Description of Solutions

Concentration Term	Definition	Examples
Molarity (M)	Moles solute per 1 L **SOLUTION** $M = (moles/L)$	3 M NaCl; Dissolve 175 g NaCl in enough water to make 1 L solution.
Osmolarity	Molarity times the number of particles (i) produced by each mole of solute Osmolarity = M x i	0.1 M NaCl; Osmolarity = 0.2 $NaCl \rightarrow Na^+ + Cl^-$; i = 2 0.3 M $Mg_3(PO_4)_2$; i = 5 Osmolarity = 1.5 0.1 M $Mg_3(PO_4)_2 \rightarrow 3\ Mg^{+2} + 2\ PO_4^{3-}$
	Osmolarity = i x M (i = 5)	Osmolarity = 0.5 0.8 M glucose (i = 1) Osmolarity = 0.8
(% w/v)	Weight solute per 100 mL **SOLUTION**	5 g of X in 100 <u>mL</u> solution 5 % w/v
(% w/w)	Weight solute per weight	5 g of X in 50 g solution

SOLUTION		10 % w/w
(% v/v)	In a 2 component solution, volume of A added to enough liquid B to make 100 mL **SOLUTION**	17 mL liquid A plus enough liquid B to make 100 mL solution 17 % v/v

SELF-TEST QUESTIONS

<u>MULTIPLE CHOICE</u>. In the following exercises, select the correct answer from the choices listed. In some cases, two or more choices will be correct.

1. Which of the following gases can dissolve in rain droplets to produce acid rain.

   a. O_2
   b. SO_3
   c. CO_2
   d. CO

2. The solubility of a solid solute in water will depend on what factors?

   a. temperature
   b. size of the solute particles
   c. polarity of solute
   d. pressure

3. Which of the following covalent molecules can hydrogen-bond to water?

   a. nitrogen, N_2
   b. ammonia, NH_3
   c. methane, CH_4
   d. ethyl alcohol, C_2H_5OH

4. Indicate which of the following properties a colloidal dispersion has that a true solution does not.

   a. homogeneity
   b. Tyndall effect
   c. filterable with ordinary paper
   d. electrical conductance

5. An example of a colloid in which a solid is dispersed in a liquid is:

   a. cheese
   b. whipped cream
   c. jellies
   d. clouds

6. Which of the following properties does a suspension **not** have?

   a. settles on standing
   b. homogeneous
   c. translucent or opaque
   d. particles size of < 1000 nm

83

7. A solution containing proteins and NaCl salt is placed in a semipermeable membrane bag and dialyzed against distilled water. Which component or components will be found in the distilled water?

    a. NaCl, proteins   c. proteins
    b. NaCl   d. none of the components

8. The solubility of a polar gas molecule in a liquid solvent increases with:

    a. increasing temperature   c. increased gas pressure
    b. decreasing temperature   d. increased solvent polarity

9. The preparation of 300 mL of a 2.0×10^{-5} M solution of hemoglobin (molecular weight = 64,000) requires how many grams of hemoglobin?

    a. 0.38 g   c. 6.4×10^{-1} g
    b. 38 g   d. 7.6 mg

10. The reaction of oxygen gas with hemoglobin (Hb) is shown in the equation below:

    $$Hb + 4O_2 = Hb(O_2)_4$$

    How many moles of O_2 can react with the hemoglobin solution in problem 9?

    a. 3×10^{-2} moles   c. 6×10^{-6} moles
    b. 2.4×10^{-5} moles   d. 3×10^{-6} moles

11. A solution of HCl is prepared to be 1×10^{-2} M. A 15 mL aliquot is taken and diluted with 80 mL of water. The final concentration of HCl is

    a. 1.9×10^{-4} M   c. 1.6×10^{-3} M
    b. 1.9×10^{-3} M   d. 1.6×10^{-4} M

12. A 1.5 M solution is made by dissolving 50 g of compound X in 50 mL of solution. The molecular weight of X is :

    a. 660   c. 150
    b. 100   d. 300

13. To make a dye solution for coloring an old white shirt, a student dissolves 5 mg of dye to 10 L of water. Calculate the concentration of this solution in ppb. (Recall that 1 mL of water weighs 1 g).

    a. 5000 ppb   c. 500 ppb
    b. 50 ppb   d. 5 ppb

84

14. The relative solubility of gases in water is:

    a. $O_2 > N_2 > CO_2 > SO_2$ b. $SO_2 > CO_2 > O_2 > N_2$
    c. $CO_2 > O_2 > N_2 > SO_2$ d. $O_2 > SO_2 > CO_2 > N_2$

15. Two liquids that are miscible are:

    a.   Gasoline & water
    b.   Oil and vinegar
    c.   water and alcohol
    d.   oil paint & water

16. Indicate which of the following is not a homogeneous solution.

    a.   Granite
    b. CO_2 and H_2O
    c.   alcohol and water
    d.   Brass (Cu/Zn)

17. A student has 300 mL of a 2 M solution of HCl. He wants to use all of this solution and made a 0.5 M solution of HCl. Calculate how much water must he add to accomplish this.

    a.   900 mL   c.   1.2 L
    b.   500 mL   d.   400 mL

18. A solution that contains 2 moles of X in 1000 g of water decreases the freezing point of water to $-11.1^0 C$. The characteristics of X are:

    a.   an electrolyte, with $i = 1$ c.   an electrolyte, with $i = 3$
    b.   an electrolyte, with $i = 2$ d.   a nonelectrolyte

19. A 1×10^{-6} M solution of X (MW = 100) contains:

    a.   100 ppm   c.   1 ppm
    b.   10 ppm   d.   0.1 ppm

COMPLETION. Write the word, phrase or number in the blank space or draw the appropriate structure in answering the question.

1. Preparation of 500 mL of a 0.5 M NaCl solution requires _____ NaCl.

2. The osmolarity of a $0.1 \underline{M}$ $Na_3(PO_4)$ solution is _____. The freezing point of this solution is determined on a 50 mL portion (aliquot) of the solution. The observed freezing point will be _____.

3. Crystalline substances that contain no water are called _____. Substances which become hydrated on exposure to the open air are referred to as _____ compounds.

4. Three 0.1 M solutions are prepared, one with HF, another with NaCl and the last one with sucrose (common sugar). Because the labels came off the bottles, we shall identify the solutions as simply solutions A, B and C. Solutions marked A and B conduct electricity; however, solution A clearly has a much greater capacity to conduct electricity than does solution B. Solution C does not conduct electricity. Identify each solution and characterize it according to its electrolytic character.

Compound	**Solution**	**Type of Electrolyte**
HF	(i)	(iv)
NaCl	(ii)	(v)
Sucrose	(iii)	(vi)

5. A true solution contains particles with a maximum diameter of _____.

6. A _____ solution has a greater osmolarity than do red blood cells. Therefore, placing red blood cells in this solution will cause the cells to _____. A 0.3 $\underline{M}$ glucose solution is a _____ (tonic) solution.

7. The concentration of a 0.40 $\underline{M}$ glucose, $C_6H_{12}O_6$, solution, can be expressed in terms of (% w/v). The equivalent concentration expressed in this way would be _____ (% w/v).

8. The random motion of a colloid suspended in a solvent is called _____.

9. A biochemist prepares a solution containing hemoglobin, the compound (a protein) in the red blood cells that carries O_2 to our cells. The hemoglobin in this solution is referred to as the _____.

10. A student adds benzene (C_6H_6) to a bottle containing CCl_4 and also to a separate bottle containing methyl alcohol, CH_3OH. The benzene is miscible in the _____, but immiscible in the _____. This is an example of the rule that "like dissolves like." In this case, a _____ compound dissolves in a _____ solvent.

11. Osmosis involves the passage of _____ from the _____ to the _____ side of a semipermeable membrane.

12. _____ is the process that is used in hospitals to remove waste products from the blood of patients who have kidneys that are not functioning properly.

13. The size of a solute particle in a solution distinguishes the solution as a true solution, a colloid or a suspension. The size of these particles are _____ for a solution, _____ for a colloid solution and _____ for a suspension.

Indicate whether the statements are true or false. The number of the question refers to the section of the book that the question was taken from.

6.2 What Are the Most Common Types of Solutions?
(a) A solute is the substance dissolved in a solvent to form a solution.
(b) A solvent is the medium in which a solute is dissolved to form a solution.
(c) Some solutions can be separated into their components by filtration.
(d) Acid rain is a solution.

6.3 What Are the Distinguishing Characteristics of Solutions?
(a) Solubility is a physical property like melting point and boiling point.
(b) All solutions are transparent—that is, you can see through them.
(c) Most solutions can be separated into their components by physical methods such as distillation and chromatography.

6.4 What Factors Affect Solubility?
(a) Water is a good solvent for ionic compounds because water is a polar liquid.
(b) Small covalent compounds dissolve in water if they can form hydrogen bonds with water molecules.
(c) The solubility of ionic compounds in water generally increases as temperature increases.
(d) The solubility of gases in liquids generally increases as temperature increases.
(e) Pressure has little effect on the solubility of liquids in liquids.
(f) Pressure has a major effect on the solubility of gases in liquids.
(g) In general, the greater the pressure of a gas over water, the greater the solubility of the gas in water.
(h) Oxygen, O_2, is insoluble in water.

6.6 Why Is Water Such a Good Solvent?
(a) The properties that make water a good solvent are its polarity and its capacity for hydrogen bonding.
(b) When ionic compounds dissolve in water, their ions become solvated by water molecules.
(c) The term "water of hydration" refers to the number of water molecules that surround an ion in aqueous solution.
(d) The term "anhydrous" means "without water."
(e) An electrolyte is a substance that dissolves in water to give a solution that conducts electricity.
(f) In a solution that conducts electricity, cations migrate toward the cathode and anions migrate toward the anode.
(g) Ions must be present in a solution for the solution to conduct electricity.
(h) Distilled water is a nonelectrolyte.
(i) A strong electrolyte is a substance that dissociates completely into ions in aqueous solution.
(j) All compounds that dissolve in water are electrolytes.

6.7 What Are Colloids?
(a) A colloid is a state of matter intermediate between a solution and a suspension, in which particles are large enough to scatter light but too small to settle out from solution.

(b) Colloidal solutions appear cloudy because the colloidal particles are large enough to scatter visible light.

ANSWERS TO SELF-TEST QUESTIONS

MULTIPLE CHOICE

1.	b, c	10.	b
2.	a, c	11.	c
3.	b, d	12.	a
4.	b	13.	c
5.	c	14.	b
6.	b, d	15.	c
7.	b	16.	a
8.	b, c, d	17.	a
9.	a	18.	c
		19.	d

COMPLETION

1. 14.6 g
2. 0.4, -0.76°C
3. anhydrous; hygroscopic
4. (i) Solution B   (iv) weak electrolyte
   (ii) Solution A   (v) strong electrolyte
   (iii) Solution C   (vi) non-electrolyte
5. 1 nm
6. hypertonic, shrivel; isotonic
7. 7.2 (% w/v)
8. Brownian motion
9. solute
10. CCl_4, CH_3OH, non-polar, non-polar
11. solvent (molecules), dilute; (more) concentrated
12. Hemodialysis
13. about 1 nm, about 1-1000 nm, greater than 1000 nm

QUICK QUIZ

6.2
    (a) True   (b) True   (c) False   (d) True

6.3
    (a) True   (b) False   (c) True

6.4
    (a) True   (b) True   (c) True   (d) False
    (e) True   (f) True   (g) True   (h) False

6.6

(a) True	(b) True	(c) False	(d) True
(e) True	(f) True	(g) True	(h) True
(i) True	(j) False		

6.7

(a) True	(b) True

Hofstadter's Law:
It always takes longer than you expect, even
when you take into account Hofstadter's Law.
   - Douglas Hofstadter

7

Reaction Rates and Chemical Equilibrium

CHAPTER OBJECTIVES

After you have studied the chapter and worked the assigned exercises in the text and the study guide, you should be able to do the following.

1. Given the concentration of the reactants and/or products at various times during the reaction, calculate the rate of reaction.

2. Distinguish between molecular collisions and effective molecular collisions and how these relate to the rate of a reaction.

3. Clarify how effective collisions, temperature and activation energy influence the rate of reaction.

4. Draw and completely label the energy diagrams for (i) an endothermic reaction and (ii) an exothermic reaction.

5. Describe how a catalyst influences a reaction and its affect on the energy diagram.

6. Define the composition in the transition state (TS) and the position of the TS in the energy diagram.

7. Discuss how each of the following factors affects the rate of reaction: (1) nature of reactants,(2) concentration of reactants and products, (3) temperature and (4) the presence of a catalyst.

8. Describe the concept of dynamic equilibrium as it relates to a reversible reaction.

9. Given the equilibrium expression, write out the chemical reaction.

10. Given the chemical reaction, write out the equilibrium expression.

11. Given the concentration of the reactants and products at equilibrium, calculate the equilibrium constant, K.

12. Use Le Chatelier's Principle to explain how the following factors influence the equilibrium: (1) addition of a reaction component; (2) removal of a reaction component; (3) addition of a catalyst.

13. Utilizing Le Chatelier's Principle and considering both an exothermic and endothermic reaction, explain how a change in temperature (increase or decrease) will alter the equilibrium and therefore the equilibrium constant.

14. Define the important terms and comparisons in this chapter and give specific examples where appropriate.

IMPORTANT TERMS AND COMPARISONS

Chemical Kinetics
Initial Reaction Rate
Effective and Ineffective Collision
Activation Energy and Energy of Reaction
Transition State and Activated Complex
Energy Diagram
Exothermic and Endothermic Reactions
Le Chatelier's Principle

Catalyst and Enzyme
Reversible and Irreversible Reaction
Dynamic Equilibrium
Haber Process
Fixation of Nitrogen
Equilibrium Constant and Equilibrium
  Expression
Rate Constant and Equilibrium Constant

SELF-TEST QUESTIONS

MULTIPLE CHOICE. In the following exercises, select the correct answer from the choices listed. In some cases, more than one choice will be correct.

1. Consider the reaction, $A + B \rightarrow C$. It was found that at 18 minutes after the start of the reaction, the concentration of C was 0.04 M. The rate of this reaction is:

    a. 450 M/min
    b. 4.2×10^{-3} M/min

    c. 900 mole/L-min
    d. 2.1×10^{-3} mole/L-min

2. Effective molecular collisions in the reaction, $A + B \rightarrow C$, are necessary for a reaction to take place because:

   a. Reactant molecules, A and B, must directly interact with each other to produce product.
   b. The collisions of rapidly moving molecules provide energy to break bonds.
   c. Both a & b.

3. The activated complex in the reaction $A + B \rightarrow C + D$ is composed of:

   a. A and B   c. A
   b. C and D   d. A and D

4. Reaction $X + Y \rightarrow Z$ is completed in 2 hours at 20°C. The reaction is carried out at a different temperature at which the reaction is complete in 7.5 minutes. This new temperature is:

   a. 293 K   c. 333 K
   b. 40°C   d. 30°C

5. The equilibrium constant is equal to 5×10^{-4} for the reaction, $A + B + C = 2D$. The equilibrium expression for the reaction is:

   a. $K = \dfrac{(A)\ (B)\ (C)}{(D)^2}$ c. $K = \dfrac{(D)^2}{(A)\ (B)\ (C)}$

   b. $K = \dfrac{(2D)}{(A)\ (B)\ (C)}$ d. $K = \dfrac{(2D)^2}{(A)\ (B)\ (C)}$

6. If the equilibrium concentrations for the reactants and products in the reaction in question 5 are $[A] = [B] = [C] = 1 \times 10^{-3}$ M, the concentration of D would be:

   a. 5×10^{-10} $\underline{M}$ c. 7×10^{-7} $\underline{M}$
   b. 5×10^{-13} $\underline{M}$ d. 5×10^{-5} $\underline{M}$

7. Consider the reaction, $Cl_{2(g)} + H_2O_{(l)} = H^+ + Cl^- + HOCl$. The molarity values for of the chemical species at equilibrium are determined to be: $[Cl_{2(g)}] = 0.10$; $[H_2O] = 0.24$; $[H^+] = 1 \times 10^{-3} = [Cl^-]$; $[HOCl] = 0.4$. The equilibrium constant for this reaction is:

   a. 1.7×10^{-5} c. 6×10^4
   b. 1.7×10^{-2} d. 6×10^2

8. Which of the following describe the Haber process?

   a. $N_{2(g)} + 3H_{2(g)} = 2NH_{3(g)}$ c. Uses a catalyst
   b. Exothermic   d. All of the above

9. If one monitors the rate of the formation of C in the reaction, $A + B \rightarrow C$, one finds the rate is 5×10^{-2} M/min. After one hour of reaction, the concentration of C is:

   a. 5×10^{-2} M   c. 8×10^{-4} M
   b. 3 M   d. 30 M

10. Indicate the statement that correctly describes the relationship between the equilibrium constant, K, and the rate of the reaction.

   a. The larger the K value, the faster the rate of the reaction is.
   b. The larger the K value, the smaller the rate of reaction is.
   c. There is no simple relationship between K and the rate of reaction.
   d. The faster the initial rate of the reaction, the larger the K value.

<u>COMPLETION</u>. Write the word, phrase or number in the blank space that completes the statement.

1. Indicate whether the following reactions are endothermic or exothermic.

   a. $A + B + \text{Energy} \rightarrow C$ _____

   b. $A + B \rightarrow C + D + \text{Energy}$ _____

   c. $A + B \rightarrow C$
      $\Delta E = -40$ kcal/mole   _____

   d. $H_{2(g)} + O_{2(g)} \rightarrow H_2O_{(e)} + \text{Energy}$ _____

2. Indicate how each of the following will affect the rate of formation of C in the reaction, $A + B = C$. The answer should be increase, decrease or no effect (on rate).

   a. Add a catalyst   _____
   b. Increase concentration of A   _____
   c. Decrease concentration of B   _____
   d. Decrease the temperature   _____

3. Consider the exothermic reaction, $A + B \rightarrow C + D$, -10 kcal, in which the activation energy for the reaction is 25 kcal. <u>Complete</u> the diagram below with labels and showing all items of importance.

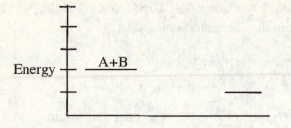

Energy ⊢ ___A+B___

⊢

⊢   ___

4. Completely label the following diagram by placing the appropriate label in each indicated block.

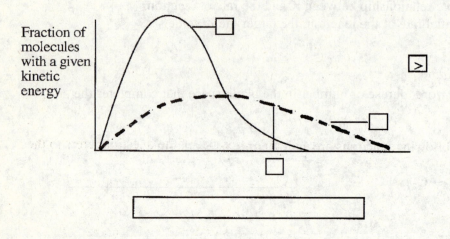

Fraction of molecules with a given kinetic energy

5. Molecules that act as catalysts in the human body are called _____.

6. Catalysts in a reaction lower the _____ of the reaction by increasing the number of _____ between reactant molecules, without increasing the reaction temperature. However, the energy of the reactants and products (do or do not) _____ change by the presence of a catalyst.

7. Complete the following table.

Equilibrium Expression   Reaction

a. $H_2 + O_2 = H_2O$

b. $\dfrac{(W)^2}{(A)\,(B)} = K$

c. $H_2O = H^+ + OH^-$

8. The concept of a dynamic equilibrium brings out the finding that although the _____ of both reactant and product molecules (does or does not) change,

both the _____ and _____ reactions are taking place.

9. Using Le Chatelier's Principle, indicate how the following changes will affect the equilibrium in the exothermic reaction. Indicate whether the reaction goes to the right or to the left or has no change.

$$A + B = C$$

Change   Effect

a. Increase [A]

b. Decrease [C]

c. Add a catalyst

d. Increase temperature

QUICK QUIZ

Indicate whether the statements are true or false. The number of the question refers to the section of the book that the question was taken from.

7.1 How Do We Measure Reaction Rates?
(a) The rate of a chemical reaction is most commonly expressed as the change in the concentration of a reactant or a product in the units of moles per liter per minute.
(b) The rate of most chemical reactions remains constant over the duration of the reaction.
(c) The reaction rate is fastest at the start of a reaction and thereafter decreases until it falls to zero.
(d) For the following reaction

$$2N_2O_5(g)) \longrightarrow 4NO_2(g) + O_2(g)$$

if the rate of appearance of $O_2(g)$ is 0.023 mol L^{-1} s^{-1}, the rate of appearance of NO_2 is 0.092 mol L^{-1} s^{-1}.

7.2 Why Do Some Molecular Collisions Result in Reaction Whereas Others Do Not?
(a) For molecules to react, they must first collide.
(b) For a collision of molecules to be effective and lead to a chemical reaction (change), the two molecules must collide with (1) sufficient energy to overcome the activation energy barrier and (2) collide in certain orientations such that the proper bonds break and new ones form.
(c) All molecular collisions result in chemical reaction.
(d) The rate of a chemical reaction is proportional to the number of effective collisions.

7.3 What Is the Relationship Between Activation Energy and Reaction Rate?
(a) The rate of a chemical reaction generally increases when the concentrations of the reactants are increased.
(b) An exothermic reaction is one in which the energy of the products is greater than that of the reactants.

95

(c) The rate of a chemical reaction depends on its activation energy; the greater the activation energy, the slower the rate of reaction.

(d) The maximum on an energy diagram represents a transition state.

(e) Activation energy is the difference in the energy between the transition state and the reactants.

7.4 How Can We Change the Rate of A Chemical Reaction?

(a) A catalyst increases the rate of formation of product(s) by providing an alternative pathway,

with a lower activation energy, by which the reaction can take place.

(b) For reactions taking place in the gas phase, increasing the pressure on the reacting system generally results in an increase in the rate of the chemical reaction.

(c) Increasing the temperature of molecules in the gas phase increases their kinetic energy (molecular velocities).

(d) In general, increasing the temperature of a reacting system results in an increase in the rate of a chemical reaction.

(e) For the reaction

$$NO(g)) + O_3(g) \longrightarrow NO_2(g) + O_2(g)$$

the rate law is

$$Rate = k[NO(g)][O_3(g)]$$

If the concentration of $NO(g)$ is doubled, the rate of reaction will double.

(f) If the concentrations of $NO(g)$ and $O_3(g)$ in (e) are both doubled, the reaction rate will increase by a factor of 4.

(g) The value of k, the rate constant for a reaction, increases as temperature increases.

7.5 What Does It Mean to Say That a Reaction Has Reached Equilibrium?

(a) After a chemical reaction reaches equilibrium, there is no further change in the concentration of either reactants or products.

(b) After a reaction reaches equilibrium, no further chemical reaction takes place.

(c) At chemical equilibrium, the rates of the forward and reverse reactions are equal.

7.6 What Is an Equilibrium Constant and How Do We Use It?

(a) From the value of an equilibrium constant, we can deduce whether products or reactants predominate at equilibrium.

(b) A value of less than 1.0 for an equilibrium constant indicates that the equilibrium position lies more on the side of products than on the side of reactants.

(c) For a reaction at equilibrium, increasing the concentrations of the reactants increases the value of the equilibrium constant for the reaction.

(d) The value of an equilibrium constant for a reaction is independent of temperature.

7.8 What Is Le Chatelier's Principle?

(a) Le Chatelier's principle states the following: If a system is at equilibrium and the conditions are changed so that it is no longer at equilibrium, the system will react to reach a new equilibrium in a way that partially counteracts the change.

96

(b) If the concentration of a reactant is increased, the system will react in the forward direction.

(c) If the concentration of a product is increased, the system will react in the reverse direction.

(d) Consider the following equilibrium.:

$$N_2O_4(g) \rightleftharpoons 2NO_2(g)$$

An increase in pressure on the reacting system will result in a decrease in the concentration of $NO_2(g)$ and an increase in the concentration of N_2O_4 (g).

ANSWERS TO SELF-TEST QUESTIONS

MULTIPLE CHOICE

1. d	4. c
2. c	5. c
3. a and b; note that A and B are the only reactants and C and D are the only products. The atoms in A and B are the same ones in C and D, only rearranged to form the new molecules.	6. c
	7. a
	8. d
	9. b
	10. c

COMPLETION

1.	a.	endothermic	c.	exothermic
	b.	exothermic	d.	exothermic
2.	a.	increase rate	c.	decrease rate
	b.	increase rate	d.	decrease rate (number of effective collisions is reduced

3.

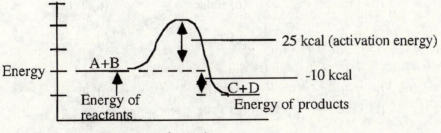

transition state

Energy

A+B

Energy of reactants

C+D

Energy of products

25 kcal (activation energy)

-10 kcal

Progress of reaction

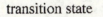

97

4.

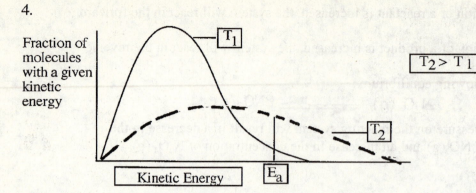

Fraction of molecules with a given kinetic energy

T_1

$T_2 > T_1$

T_2

Kinetic Energy

E_a

5. enzymes
6. activation energy, effective collisions, do not
7. a. $K = \dfrac{(H_2O)}{(H_2)(O_2)}$

   b. $A + B = 2 W$

   c. $K = \dfrac{(H^+)(OH^-)}{(H_2O)}$

8. concentrations, does not, forward, reverse
9. a. reaction goes to right
   b. reaction goes to right
   c. no effect
   d. reaction goes to left. In an exothermic reaction, heat can be considered a product. Therefore, increasing the temperature will increase the heat (a product) and the reaction will go to the left to relieve the stress.

QUICK QUIZ

7.1
| (a) True | (b) False | (c) True | (d) True |

7.2
| (a) True | (b) True | (c) False | (d) True |

7.3
| (a) True | (b) False | (c) True | (d) True |
| (e) True | | | |

7.4
| (a) True | (b) True | (c) True | (d) True |
| (e) True | (f) True | (g) True | |

7.5
| (a) True | (b) False | (c) True | |

7.6
| (a) True | (b) False | (c) False | (d) False |

7.8
| (a) True | (b) True | (c) True | (d) True |

Concerning solutions to acid-base
equilibrium problems: The answers
to the riddle of life are not found
in the back of the book.
   - Charlie Brown

Acids and Bases

CHAPTER OBJECTIVES

As will become evident in this chapter and those focused on biochemistry (Chapters 19-31), the body contains aqueous systems vital to life as we know it. These systems include intracellular fluids, blood, gastric juices, saliva and more. The importance of understanding the **CHARACTERISTICS AND REACTIONS** of **ACIDS** and **BASES** in aqueous solutions therefore cannot be overemphasized. This chapter dwells on these fundamentals.

After you have studied the chapter and worked the assigned exercises in the text and the study guide, you should be able to do the following.

1. Characterize Arrhenius acids and bases and define the similarities and differences with Bronsted-Lowry acids and bases. Give specific examples of each.

2. Explain the difference between strong and weak acids (bases) and give four examples of strong acids (bases).

3. Write the reactions involving Bronsted-Lowry acids and bases and point out the conjugate acid-base pairs.

4. Define Ka and pKa values and explain how they are related to acid strength.

5. Write ionic equations for the neutralization reaction of acids and bases; clearly identify the salt that forms in each case.

6. For an aqueous solution, define the neutral point and the acidic and basic domains in terms of pH, pOH, [H$^+$] and [OH$^-$]. At any pH, show the relationship between pH and pOH and also [H$^+$] and [OH$^-$].

7. Given the [H$^+$] or [OH$^-$] of a solution, calculate the pH. Given the pH or pOH of a solution, calculate the [H$^+$].

8. Name six solutions that are body fluids, drinks, or common household products which are (1) acidic or which are (ii) basic. In each case, indicate its approximate pH value.

9. Explain the importance of acid-base indicators in qualitatively testing the acidic or basic character of a solution.

10. Define buffer solution and indicate the components and their relative concentrations necessary to prepare buffers with good buffering capacity.

11. Knowing the Ka or the pKa of a weak acid, indicate what the components of the buffer are and the ideal ratio of components that are needed to produce an effective buffer solution.

12. Indicate the components of a buffer that react with any added acid and what components are available to react with any added base.

13. Indicate the characteristics of the buffers developed by Good (called "Good" buffers).

14. Show how the components of carbonate and phosphate buffers react on addition of strong acid or base.

15. Explain in some detail the way in which you would determine the concentration of a known volume of basic solution by titration with a strong acid of known concentration.

16. Name a few acids that are not monoprotic and have more than one titratable proton - referred to as diprotic and triprotic acids. Explain how this must be taken into account in calculations for an acid-base titration.

17. Write out the relationship in the Henderson-Hasselbalch equation and indicate types of problems in which this is used.

18. Define the important terms and comparisons in this chapter and give specific examples where appropriate.

IMPORTANT TERMS AND COMPARISONS

Arrhenius Acid and Base	Ion Product of Water, Kw
Hydronium Ion and Hydroxide Ion	Acidic and Basic Solution
Strong and Weak Acid	Amphiprotic Substance
Mono-, Di- and Triprotic Acids	pH Indicator and Color Change

Glacial Acetic Acid
Bronsted-Lowry Acid and Base
Acid Ionization Constant, Ka
Acid-Base Titration
pKa, pH and pOH
Buffer Solutions
Hyper- and Hypoventilation
Titration, End Point and Equivalence Point
Henderson-Hasselbalch Equation
Good Buffers - TRIS, HEPES and MOPS

pH Paper and pH Meter
Conjugate Acid-Base Pair
Metals, Metal Oxides and Metal Hydroxides
Buret and Pipet
Ionic Equation
Buffering Capacity
Antacid
Baking Soda
Acidosis and Alkalosis
Zwitterions

SELF-TEST QUESTIONS

MULTIPLE CHOICE. In the following exercises, select the correct answer from the choices listed. In some cases, more than one choice will be correct.

1. The pH of a 1×10^{-4} M NaOH solution is

   a. 4
   b. 1

   c. 10
   d. 6

2. Which of the following is (are) strong acid(s)?

   a. perchloric acid
   b. boric acid

   c. hydroiodic acid
   d. acetic acid

3. Which of the following is both a strong acid and an oxidizing agent?

   a. NH_3
   b. H_2SO_4

   c. HNO_3
   d. H_2O

4. Indicate which of the following characteristics is (are) inherently associated with a strong acid solution.

   a. concentrated
   b. low pH value

   c. dissociates close to 100%
   d. reacts with bases

5. The species in a phosphate buffer at pH 6.5 that reacts with added hydroxide ion is:

   a. $H_2PO_4^-$
   b. H_3PO_4

   c. HPO_4^{-2}
   d. PO_4^{-3}

6. A 100 mL solution of 0.1 M phosphoric acid is titrated with a 0.1 M NaOH solution. To reach the equivalence point will require the addition of what volume of NaOH?:

    a. 100 mL   c. 300 mL
    b. 33.3 mL   d. 200 mL

7. The Ka value for lactic acid is 1.4×10^{-4}. The pKa value is:

    a. 4.1   c. 4.6
    b. 3.8   d. 3.1

8. A solution with pH = 7.7 has a $[H^+]$ of:

    a. 2×10^{-7} M   c. 2×10^{-8} M
    b. 7×10^{-7} M   d. 7×10^{-8} M

9. The pKa values for four acids are 2.6, 3.6, 4.9 and 5.2. The weakest acid is the one with pKa value of:

    a. 2.6   c. 4.9
    b. 3.6   d. 5.2

10. The addition of base to a solution at pH =5 changes the pH to pH = 1. The concentration of H^+ increases by:

    a. 4   c. 100
    b. 10   d. 10,000

11. The addition of strong acid to a solution that is initially at pH 8 does not change the pH. The original solution:

    a. contains a strong acid   c. is a buffer solution
    b. contains an indicator   d. is past its equivalence point

12. Consider the reaction: $PO4^{-3} + H2S = HPO4^{-2} + HS^-$
    Refer to Table 8.2 in the text for a listing of acids and their conjugate bases.
    The position of the equilibrium;

    a. lies toward the reactants
    b. lies toward the products
    c. cannot be predicted
    d. will have n equilibrium constant of K =1

13. The pOH of a solution made up of 2×10^{-4} moles of $Mg(OH)_2$ in 400 mL of water is:

    a. 11   c. 3
    b. 12   d. 2

102

14. Considering the acids below, together with the corresponding conjugate bases, which buffer system would be used to produce a buffer solution at pH 9.0?

    a.   acetic acid, $K_a = 1.8 \times 10^{-5}$

    b.   boric acid, $K_a = 7.3 \times 10^{-10}$

    c.   formic acid, $K_a = 1.8 \times 10^{-4}$

    d.   carbonic acid, $K_a = 4.3 \times 10^{-7}$

15. The Henderson-Hasselbalch equation can be used to calculate which of the following:
    a.   whether an acid is strong or weak.
    b.   whether a reaction between an acid and base will go to completion.
    c.   the extent of acidosis or alkalosis in the blood.
    d.   the ratio of the concentrations of the acid and its conjugate base to be used in preparing a buffer solution at the desired pH.

16. A buffer may be prepared by mixing equal amounts of a

    a.   weak acid and its conjugate base   c.   strong acid and its conjugate base
    b.   weak acid and NaOH   d.   strong acid and a weak base

17. A solution of acetic acid (pK = 4.75) is titrated with NaOH until the ratio of [acetate]/[acetic acid] = 10. Using the Henderson Hasselbalch, calculate the pH of the solution.

    a.   3.75   c.   4.75
    b.   5.75   d.   5.00

18. The "p" in front of pH and pK, or anything, indicates the:

    a.   log   c.   antilog
    b.   -log   d.   exponent

COMPLETION. Write the word, phrase or number in the blank space in answering the question.

1. Complete and balance the following reactions.

    a. $HCl + NaOH \rightarrow$ _____ + _____ + _____

    b. $Na_2O + HNO_3 \rightarrow$ _____ + _____ + _____

    c. $H_2SO_4 +$ _____ $\rightarrow K_2SO_4 + 2H_2O$

    d. $NH_3 + HCl \rightarrow Cl^- +$ _____

e. $Na_2CO_3 + HI \rightarrow$ _____ + _____ + _____
final products

2.   A 100 mL sample of unknown acid is titrated with 0.42 M NaOH. The end-point of the titration requires 29 mL of base. The molarity of the acid is _____.

3.   If the acid in question 2 were H_2SO_4, the molarity of the H_2SO_4 would be

_____.

4.   Indicate (yes or no) whether the following bases are Arrhenius and/or Bronsted-Lowry bases.

	Arrhenius	Bronsted-Lowry
a. NH_3		
b. NaOH		
c. $Al(OH)_3$		

5.   The term hydronium ion is used interchangeably with _____ , _____
and _____.

6.   The _____ of an acid in an intrinsic property of an acid, while the concentration is a variable and depends on the preparation of the solution.

7.   Indicate the conjugate acid or base to the species below.

a. H_3O^+ _____ (conjugate base)

b. H_2O _____ (conjugate base)

c. HSO_4^- _____ (conjugate acid)

d. Cl^- _____ (conjugate acid)

8.   Indicate whether the acids shown below are mono-, di- or triprotic acids.

a. H_3PO_4 _____

b. CH_3COOH _____

c. H_2S _____

d. H_2CO_3 _____

e. $H_2PO_4^-$ _____

9. Complete the following table

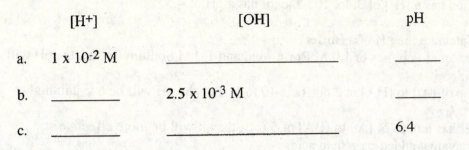

	$[H^+]$	$[OH]$	pH
a.	1×10^{-2} M	_____	_____
b.	_____	2.5×10^{-3} M	_____
c.	_____	_____	6.4

10. Using Table 8.2 in the text, indicate whether the following reactions will go to the right (to products) or to the left (to reactants).

  a. $HCN + H_2O = CN^- + H_3O^+$

  b. $H_2SO_4 + NH_3 = HSO_4^- + NH_4^+$

  c. $C_2H_5OH + Cl^- = C_2H_5O^- + HCl$

11. If the pH of the blood goes below the normal pH, the condition is called _____. This may be caused by _____, a difficulty in breathing.

QUICK QUIZ
    Indicate whether the statements are true or false. The number of the question refers to the section of the book that the question was taken from.

8.2 How Do We Define the Strength of Acids and Bases?
    (a) A strong acid is an acid at a high concentration.
    (b) A strong acid is one that dissociates completely in aqueous solution.
    (c) HCl, HNO_3, and CH_3COOH are all strong acids.
    (d) When acetic acid, CH_3COOH, is dissolved in water, most of the molecules are found as CH_3COO^-.
    (e) Acetic acid is considered a weak acid if its concentration is 0.01 M or less, and a strong acid if its concentration is greater than 5 M.

8.5 How Do We Use Acid Ionization Constants?
    (a) Phosphoric acid is a strong acid.
    (b) Carbonic acid is a stronger acid than phenol.
    (c) Phosphoric acid is the strongest of the weak acids.
    (d) The pK_a for an acid with a K_a of 3×10^{-4} is 3.52.
    (e) The K_a for an acid with a pK_a of 9.0 is 1×10^{-8}.

8.8 What are pH and pOH?
    (a) A solution is acidic if it has a pH of 6.5.
    (b) A weak acid will always have a higher pH than a strong acid.

(c) Soft drinks are usually basic.

(d) If a solution has a pH of 5, it also has a pOH of 9.

(e) If a solution has a $[H^+]$ of 3.4×10^{-5}, then it has a pH of 4.5.

8.11 How Do We Calculate the pH of a Buffer?

(a) If we mix equal volumes of 1.0 M Boric Acid and 1.0 M Sodium Borate, the pH will be 8.8.

(b) If the ratio of $[A^-]$ to $[HA]$ in a buffer is 10 to 1, then the pH will be one unit higher than the pK_a.

(c) If a buffer has a ratio of $[A^-]$ to $[HA]$ of 5 to 1, then it will be more effective at buffering against added base than acid.

(d) Boric acid would make an effective blood buffer.

ANSWERS TO SELF-TEST QUESTIONS

MULTIPLE CHOICE

1.	c	9.	d
2.	a, c	10.	d
3.	b, c	11.	c
4.	c, d	12.	b
5.	a	13.	c
6.	c	14.	b
7.	b	15.	d
8.	c	16.	a
		17.	a
		18.	b

COMPLETION

1. a. H_2O, Na^+, Cl^- d. NH_4^+

    b. H_2O, 2 Na^+, 2 NO_3^- e. H_2O, $CO_2\uparrow$, 2 NaI

    c. 2 KOH

2. 0.12 M

3. 0.06 M

4. a. no, yes

    b. yes, yes

    c. yes, yes

5. proton, H^+, H_3O^+

6. strength

7. a. H_2O c. H_2SO_4

    b. OH^- d. HCl

8. a. triprotic   d. diprotic

    b. monoprotic   e. diprotic

    c. diprotic

106

9.   a. 1×10^{-12} M, 2
      b. 4.0×10^{-12}, 11.4
      c. 4×10^{-7}, 2.5×10^{-8}

10. a.   towards reactants
      b.   toward products
      c.   toward reactants

11. acidosis, hypoventilation

QUICK QUIZ

8.2

(a) False	(b) True	(c) False	(d) False
(e) False			

8.5

(a) False	(b) True	(c) True	(d) True
(e) False			

8.8

(a) True	(b) False	(c) False	(d) True
(e) True			

8.11

(a) False	(b) True	(c) False	(d) False

Every man, woman and child inhabiting this planet
became an unwitting guinea pig in a vast
biochemical experiment on July 16, 1945, when
the first atomic bomb was successfully detonated.
- Ernest Borek
The Atoms Within Us

9

Nuclear Chemistry

CHAPTER OBJECTIVES

After you have studied the chapter and worked the assigned exercises in the text and the study guide, you should be able to do the following.

1. Characterize the three major types of radioactivity in terms of symbol, charge, mass and penetrating capability.

2. Associate the names of famous scientists with their accomplishment in nuclear chemistry.

3. Given the frequency (υ) of electromagnetic radiation and the equation $(\lambda)(\upsilon) = c$, calculate the wavelength (λ). Given the wavelength, calculate the frequency.

4. Order the different forms of radiation in the electromagnetic spectrum in terms of increasing energy and frequency and decreasing wavelength.

5. Explain the fundamental difference between a chemical and a nuclear reaction.

6. Write balanced nuclear equations for a reaction involving the emission of (i) an alpha particle, (ii) a beta particle and (iii) a gamma emission. Note whether the product nucleus has a mass greater than, equal to or less than the original radioactive nuclide.

7. Given an equation that includes all but one particle in a nuclear reaction, balance the equation and identify the missing particle.

8. Explain the term "half-life" and relate this to the decay curve for a radioactive nuclide.

9. Explain how a Geiger-Muller counter functions in monitoring the intensity of beta particle emission.

10. Define the three units, roentgens, rems and rads, used to describe the effects of radiation and explain how they relate to each other.

11. Outline how ionizing radiation damages tissue by the formation of free radicals.

12. Write out the relationship that describes how the intensity of ionizing radiation varies with distance.

13. Describe the role of cobalt-60, iodine-131, and technetium-99m in medical diagnosis and/or treatment.

14. Explain the importance of fusion reactions to life on Earth.

15. Point out the most obvious differences between nuclear fusion and fission reactions. Outline how the fission reactions can be used to produce controlled nuclear energy on the one hand and the atomic bomb on the other hand.

16. Define the important terms and comparisons in this chapter and give specific examples where appropriate.

IMPORTANT TERMS AND COMPARISONS

Alpha (α) Particle, Beta (β) Particle,
   Positron Particle and Gamma (γ) Ray
Equivalent Dose
Wavelength (λ) and Frequency (υ)
Radiation (Exposure) Badge
Photon and Electron
Nuclide
Background Radiation
Tagging a Drug
Mass Number and Atomic Number
Hydrogen, Deuterium and Tritium
Isotope
Alpha, Beta, Gamma and Positron Emitter
$E = mc^2$
Isotope Generator
Half-Life ($\tau_{1/2}$)
Curie

Roentgen, Rad and Rem
Transmutation
Electromagnetic Spectrum
$(\lambda)(\upsilon) = c$
Stable and Radioactive Isotope
Subatomic Particles
Ground State
   Hyperthyroidism
Collimated Radiation Beam
Radiation Sickness
"Big Bang" Explosion
Fusion
Ionizing Radiation
Natural and Artificial Transmutation
Scintillation Counter
Radioactive Decay Curve
Nuclear Fission and Fusion

Nuclear Power Plant   "Phosphor"
Control Rods   Geiger-Muller Counter
Nuclear Waste   Spectrum of Radiation
$(I_1)(d_1^2) = (I_2)(d_2^2)$ Magnetic Resonance Imaging (MRI)
Positron Emission Tomography (PET)
Nuclear Fission and a Self-propagating "Chain" Reaction

SELF-TEST QUESTIONS

<u>MULTIPLE CHOICE</u>. In the following exercises, select the correct answer from the choices listed. In some cases, more than one choice will be correct.

1. A 4 g sample of a radionuclide at room temperature has a half-life of three months. The amount of sample remaining after a year is:

   a.   0.25 g   c.   2 g
   b.   0.40 g   d.   4 g

2. If the radioactive decay of the sample in exercise 1 were carried out -100°C, the decay process would:

   a.   decrease   c.   remain unchanged
   b.   increase   d.   cannot determine

3. A nucleus that emits a beta particle will be transformed into a product nucleus of the:

   a.   same mass number
   b.   same atomic number
   c.   atomic number one unit greater
   d.   mass number one unit greater

4. The wavelength of blue light is 4000 A (1 A = 10^{-8} cm). Its frequency is:

   a. 12×10^{-12} s^{-1} c. 7.5×10^{10} s^{-1}
   b. 7.5×10^{14} s^{-1} d. 12×10^{10} s^{-1}

5. A scintillation counter uses a "phosphor" to:

   a.   determine the half-life of a radioactive isotope
   b.   produce transuranium elements
   c.   determine the age of a very old object
   d.   measure the radioactivity in a sample

6. A millicurie is a unit of radioactivity equivalent to:

    a. 3.7×10^{10} counts s^{-1} c. 3.7×10^7 counts s^{-1}
    b. 3.7×10^{10} counts min^{-1} d. 3.7×10^4 counts s^{-1}

7. The radiation that causes the most damage within the cell is:

    a. alpha particles   c. gamma rays
    b. beta particles   d. X-rays

8. Which of the following radioactive nuclides decay the fastest?

    a. $^{197}_{80}$Hg, $t_{1/2} = 65$ hrs   c. $^{238}_{92}$U, $t_{1/2} = 4 \times 10^9$ yr.

    b. $^{210}_{84}$Po, $t_{1/2} = 138$ days   d. $^{131}_{53}$I, $t_{1/2} = 8$ days

9. The wavelength of which radiation is the longest?

    a. microwaves   c. visible light
    b. X-rays   d. radiowaves

10. The number of neutrons in phosphorus-32, $^{32}_{15}$P is:

    a. 32   c. 17
    b. 15   d. 5

11. Which of the following nuclear reactions does not involve natural transmutation?

    a. $^{11}_{5}$B $\rightarrow$ $^{11}_{5}$B + γ

    b. $^{210}_{84}$Po $\rightarrow$ $^{206}_{82}$Pb + $^{4}_{2}$He

    c. $^{14}_{6}$C $\rightarrow$ $^{14}_{7}$N + $^{0}_{-1}$e

12. A source of ionizing radiation is found to have an intensity of 25 mCi at a distance of 1 m. To reduce the intensity of this radiation to 1 mCi, calculate what distance would be needed.

    a. 5 m   c. 2 m
    b. 25 m   d. 10 m

111

13. Radon is a very dangerous radioactive element and is thought to be responsible for many cases of lung cancer. Radon is characterized by:

    a.   it is a gas
    b.   it is a major product of radioactive decay of uranium
    c.   it has a half-life of 3.8 days
    d.   it decays to radioactive isotopes of polonium, which are solids
    e.   all of the above

COMPLETION. Write the word, phrase or number in the blank space in answering the question.

1. Consider the reactions in question 11 in the multiple choice section. Reaction a) is an example of a(n) _____ emitter, reaction b) shows a(n) _____ emitter, while reaction c) exhibits a(n) _____ emitter.

2. Complete the following nuclear reactions. Refer to a periodic table.

    a. $^{28}_{13}\text{Al} \rightarrow \beta +$ _____

    b.   _____ $\rightarrow ^{1}_{1}\text{H} + ^{0}_{-1}\text{e}$

    c. $^{254}_{102}\text{No} \rightarrow ^{4}_{2}\text{He} +$ _____

    d. $^{87}_{36}\text{Kr} \rightarrow ^{87}_{37}\text{Rb} +$ _____

3. The order of increasing penetrating power for gamma rays, alpha and beta particles is:

    _____ < _____ < _____

4. Match up the names of the famous scientist in the left-hand column with accomplishments or subjects most associated with the individual.

    a.   H. Becquerel   (i)   Discovered X-ray
    b.   E. Fermi   (ii) $E = mc^2$
    c.   G. Seaborg   (iii) MRI and imaging techniques
    d.   A. Einstein   (iv)Prepared transuranium elements
    e.   W. Roentgen   (v)   Discovered nuclear fission
    f.   P. Lauerbur and Sir P. Mansfield (vi) 3.7×10^{10} counts per second
    g.   M. Curie   (vii) Observed that certain rocks gave off mysterious radiation

112

5. Radiation that is derived from both cosmic rays and rocks in our environment contribute to the so-called _____ radiation.

6. Another name commonly used for an alpha particle is the _____ of a _____ atom.

7. A beta particle has mass number of _____ and an atomic number of _____.

8. Cosmic rays are streams of _____, while gamma rays consist of _____ _____ _____.

9. Isotopes have the same number of _____ but different number of _____.

10. A sample of a radioactive nuclide with a half-life of 10 years is left in a cave for 35 years. After this time, only 0.5 grams remains. The amount of this radioactive material originally in the cave was _____.

11. A sample of radioactive isotope has an activity of 4.0 mCi. This sample emits _____ counts per second.

12. Diagnostic X-rays are absorbed by the bones and not the _____ because the bones contain elements of _____ atomic number which absorb X-rays strongly.

13. Consider the following two reactions:

    a. $$^{252}_{98}Cf + ^{10}_{5}B \rightarrow ^{257}_{103}Lr + 5\,^{1}_{0}n$$

    b. $$^{235}_{92}U + ^{1}_{0}n \rightarrow ^{90}_{37}Rb + ^{144}_{55}Cs + 2\,^{1}_{0}n$$

    Reaction (a) is an example of a _____ reaction, while reaction (b) is a _____ reaction.

14. Producing an atomic bomb and the generation of nuclear power depend on the same reaction of a _____ atom being bombarded by a neutron .This reaction produces _____ neutrons that leads to a _____ reaction that is self-sustaining and culminates in a massive release of energy (i.e., explosion). This explosion does not occur in nuclear power plants because the reaction is _____ with the use of _____ control rods.

15. A _____ _____ is a highly reactive compound with _____ electrons.

16. Medical diagnosis has recently made significant advances in that physicians can diagnose many medical problems by non-invasive procedures (no surgical procedures). Complete the following table that characterizes three of these procedures.

113

Acronym	Complete Name	Probe (Radiation) Used In Diagnosis

PET (scan)

MRI

QUICK QUIZ
Indicate whether the statements are true or false. The number of the question refers to the section of the book that the question was taken from.

9.2 What Is Radioactivity?
(a) Radioactivity refers to radiation emitted by a nucleus, all forms of which have energies and wavelengths similar to those of radiowaves.

(b) Alpha particles are helium atoms.

(c) Beta particles are electrons emitted from certain nuclei.

(d) Gamma rays have the same charge as electrons but no mass.

(e) All electromagnetic radiation consists of waves that can be described in terms of their frequency and wavelength.

(f) Wavelength is the distance between one wave crest and the next, and is given the symbol λ.

(g) The longer the wavelength of a type of radiation, the lower its frequency.

(h) The only difference between one form of electromagnetic radiation and another is the wavelength.

(i) Radiowaves have longer wavelengths than gamma rays.

(j) UV radiation is a more energetic form of radiation than gamma rays.

9.3 What Happens When a Nucleus Emits Radioactivity?
(a) When balancing a nuclear equation, the sum of the mass numbers and the sum of the atomic numbers on each side of the equation must be the same.

(b) When a nucleus emits a beta particle, the new nucleus has the same mass number but an atomic number one unit higher.

(c) When iron-59 ($^{59}_{26}$Fe) emits a beta particle, it is converted to cobalt-59 ($^{59}_{27}$Co).

(d) When a nucleus emits a beta particle, it first captures an electron from outside the nucleus and then spits it out.

(e) For the purposes of determining atomic numbers in a nuclear equation, an electron is assumed to have a mass number of zero and an atomic number of -1.

(f) When a nucleus emits an alpha particle, the new nucleus has an atomic number two units higher and a mass number four units higher.

(g) When uranium-238 ($^{238}_{92}$U) undergoes alpha emission, the new nucleus is thorium-234 ($^{234}_{90}$Th).

(h) A positron is sometimes referred to as a positive electron.

(i) When a nucleus emits a positron, the new nucleus has the same mass number but an atomic number one unit lower.

(j) When carbon-11 ($^{11}_{6}$C) emits a positron, the new nucleus formed is boron-11 ($^{11}_{5}$B).

(k) Alpha emission and positron emission both result in the formation of a new nucleus with a lower atom number.

(l) When a nucleus emits gamma radiation, the new nucleus formed has the same mass number and the same atomic number.

(m) When a nucleus captures an extranuclear electron, the new nucleus formed has the same atomic number but a mass number one unit lower.

(n) When gallium-67 ($^{67}_{31}$Ga) undergoes electron capture, the new nucleus formed is germanium-67 ($^{67}_{32}$Ge).

9.4 What is Nuclear Half-life?

(a) Half-life is the time it takes for one-half of a radioactive sample to decay.

(b) The concept of half-life refers to nuclei undergoing alpha, beta, and positron emission; it does not apply to nuclei undergoing gamma emission.

(c) At the end of two half-lives, one-half of the original radioactive sample remains; at the end of three half-lives, one-third of the original sample remains.

(d) If the half-life of a particular radioactive sample is 12 minutes, a time of 36 minutes represents three half-lives.

(e) At the end of three half-lives, only 12.5 percent of an original radioactive sample remains.

9.5 How Do We Detect and Measure Nuclear Radiation?

(a) Ionizing radiation refers to any radiation that interacts with neutral atoms or molecules to create positive ions.

(b) Ionizing radiation creates positive ions by striking a nucleus and knocking one or more electrons from the nucleus.

(c) Ionizing radiation creates positive ions by knocking one or more extranuclear electrons from a neutral atom or molecule.

(d) The curie (Ci) and the becquerel (Bq) are both units by which we report radiation intensity.

(e) The units of a curie (Ci) are disintegrations per second (dps).

(f) A microcurie (μCi) is a smaller unit than a curie (Ci).

(g) The intensity of radiation is inversely related to the square of the distance from the radiation source; for example, the intensity at 3 m from the source is 1/9 of the intensity it is at the source.

(h) Alpha particles are the most massive and highly charged type of nuclear radiation and, therefore, are the most penetrating type of nuclear radiation.

(i) Beta particles have both a smaller mass and a smaller charge than alpha particles and, therefore, are more penetrating than alpha particles.

(j) Gamma rays, with neither mass nor charge, are the least penetrating type of nuclear radiation.

9.6 How Is Radiation Dosimetry Related to Human Health?

(a) The roentgen (R) is a measure of the energy delivered from a radiation source.

(b) Roentgens are measured by the number of positive ions created by the radiation from a radioactive source.

(c) The term "rad" is an acronym for **r**adiation **a**bsorbed **d**ose.

(d) The rad takes into account the fact that different tissues may absorb different amounts of the same delivered radiation.

(e) The term "rem" is an acronym for **r**oentgen **e**quivalent for **m**an.

(f) The rem takes into account the fact that not all types of radiation cause the same tissue damage; for example, the radiation damage from 1 rad of alpha particles may be different from the damage caused by 1 rad of gamma radiation.

(g) The most useful measure of the degree to which radiation will cause tissue damage is the equivalent dose expressed in rems.

9.7 What Is Nuclear Medicine?

(a) Of the radioisotopes listed in Table 9.5, the majority decay by beta emission.

(b) Isotopes that decay by alpha emission are rarely, if ever, used in nuclear imaging because alpha emitters are rare.

(c) Gamma emitters are widely used in medical imaging because gamma radiation is penetrating and, therefore, can easily be measured by radiation detectors outside the body.

(d) When selenium-75 ($^{75}_{34}Se$) decays by electron capture and gamma emission, the new nucleus formed is arsenic-75 ($^{75}_{33}As$).

(e) When iodine-131 ($^{131}_{53}I$) decays by beta and gamma emission, the new nucleus formed is xenon-131 ($^{131}_{54}Xe$).

(f) In positron emission tomography (PET scan), the detector counts the number of positrons emitted by a tagged material and the locations within the body where the tagged material accumulates.

(g) The use of 18-fluorodeoxyglucose (FDG) in PET scans of the brain depends on the fact that FDG behaves in the body much like glucose does.

(h) A goal of radiation therapy is to destroy pathological cells and tissues without at the same time damaging normal cells and tissues.

9.8 What Is Nuclear Fusion

(a) In nuclear fusion, two nuclei combine to form a new nucleus.

(b) The energy of the Sun is derived from the fusion of two hydrogen-1 ($^{1}_{1}H$) nuclei to form a helium-4 ($^{4}_{2}He$) nucleus.

(c) The energy of the Sun arises because, once two hydrogen nuclei fuse, the two positive charges no longer repel each other.

(d) Fusion of hydrogen nuclei in the sun results in a small decrease in mass, which appears as an equivalent amount of energy.

(e) Einstein's famous $E = mc^2$ equation refers to the energy released when two particles of the same mass collide with the speed of light.

(f) Nuclear fusion occurs only in the Sun.

(g) Nuclear fusion can be carried out in the laboratory.

9.9 What is Nuclear Fission and How Is It Related to Atomic Energy?

(a) Nuclear fission refers to the splitting of a larger nucleus into two or more smaller nuclei.

(b) The fission of uranium-235 to barium-141 and krypton-92 is a spontaneous process.

(c) The energy liberated upon nuclear fission arises because the products have slightly less mass than the starting materials.

(d) The United States leads all other countries in the percentage of its energy generated by nuclear fission.

116

ANSWERS TO SELF-TEST QUESTIONS

MULTIPLE CHOICE

1.	a	7.	a
2.	c	8.	a
3.	a, c	9.	d
4.	b	10.	c
5.	d	11.	a
6.	c	12.	a
		13.	e

COMPLETION

1. a. gamma
   b. alpha
   c. beta
2. a. $^{28}_{14}Si$
   b. $^{1}_{0}n$
   c. $^{250}_{100}Fm$
   d. $^{0}_{-1}e$
3. alpha < beta < gamma
4. a. (vii)   d. (ii)
   b. (v)   e. (i)
   c. (iv)   f. (iii)
                g. (vi)
5. background
6. nucleus, helium
7. zero, minus one
8. nuclei (mainly protons), high energy radiation
9. protons, neutrons
10. ca. 6 g
11. 1.5×10^8
12. soft tissue, higher
13. fusion, fission
14. uranium, three, chain, controlled or regulated, boron
15. free radical; unpaired
16. PET (scan)   Positron Emission Tomography   Isotopes emitting positrons
    MRI   Magnetic Resonance Imaging   Radio-frequency waves

QUICK QUIZ

9.2

(a) False	(b) False	(c) True	(d) False
(e) True	(f) True	(g) True	(h) False
(i) True	(j) False		

9.3

(a) True	(b) True	(c) True	(d) False
(e) True	(f) False	(g) True	(h) True
(i) True	(j) True	(k) True	(l) True
(m) False	(n) False		

9.4

(a) True	(b) False	(c) False	(d) True
(e) True			

9.5

(a) True	(b) False	(c) True	(d) True
(e) True	(f) True	(g) True	(h) False
(i) True	(j) False		

9.6

(a) True	(b) True	(c) True	(d) True
(e) True	(f) True	(g) True	

9.7

(a) False	(b) True	(c) True	(d) True
(e) True	(f) True	(g) True	(h) True

9.8

(a) True	(b) False	(c) False	(d) True
(e) False	(f) False	(g) True	

9.9

(a) True	(b) True	(c) True	(d) False

Gasoline - 29¢/gal
- Sign at gas station in
Forty Fort, PA; circa 1965

10

Organic Chemistry

CHAPTER OBJECTIVES

After you have studied the chapter and worked the assigned exercises in the text and the study guide, you should be able to do the following.

1. Indicate the importance of the experiment that F. Wohler carried out in which he produced urea from ammonium chloride and silver cyanate.

2. Compare and contrast the properties of organic and inorganic compounds.

3. Review the octet rule and how to draw Lewis structures for simple organic molecules, as outlined in Chapter 4.

4. Review the valence-shell electron pair repulsion model (VSEPR) in Chapter 4 and predict the structures and bond angles in simple organic molecules.

5. Be able to recognize and structurally define the different organic functional groups. The major functional groups include alcohols, amines, aldehydes, ketones, carboxylic acids and carboxylic esters.

6. Define the difference between a primary, secondary and tertiary alcohol and be able to recognize and draw each of them.

7. Distinguish between primary, secondary and tertiary amines and be able to draw each of them.

8. Recognize the difference between an aldehyde, ketone, carboxylic acid and carboxylic ester functional group, all of which contain a carbonyl group, and be able to draw each of them

9. Define the important terms and comparisons in this chapter and give specific examples where appropriate.

IMPORTANT TERMS AND COMPARISONS

Organic Chemistry
VSEPR Model
Functional Groups
Condensed Structural Formulas
Primary, Secondary and Tertiary Amines
Alcohols, Amines, Aldehydes, Ketones, Carboxylic Acids and Carboxylic Esters

Lewis Structures
Synthetic and Natural Compounds
Molecular and Structural Formula
Primary, Secondary and Tertiary Alcohols

SELF-TEST QUESTIONS

<u>MULTIPLE CHOICE</u>. In the following exercises, select the correct answer from the choices listed. In some cases, more than one choice will be correct.

1. The molecular formula, C_3H_8O, may represent the following class(es) of organic compound(s). Indicate which one is correct.

   a. aldehyde
   b. carboxylic acid

   c. amine
   d. alcohol

2. Indicate which of the following Lewis structures is correct.

   a.

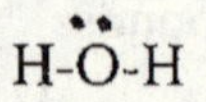

   H-Ö-H

   c.

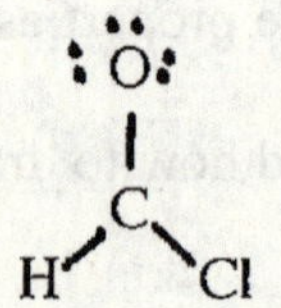

   :Ö:
   ‖
   C
   H   Cl

   b.

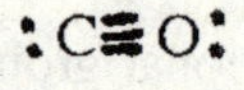

   :C≡O:

   d.

   :F̈-N̈-F̈:
        F̈

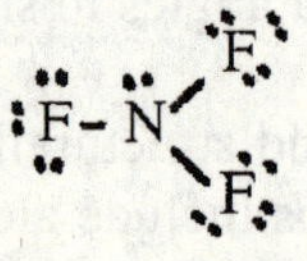

3. Indicate what type of amine the molecule, ethylamine, is.   CH3-CH2-N-H
                                                                          |
                                                                          H

   a. primary amine
   b. secondary amine

   c. tertiary amine
   d. quaternary amine

4. A tertiary carbon is bonded directly to:

   a. 2 hydrogens
   b. 3 carbons

   c. 2 carbons
   d. 4 carbons

5. The angle between the carbon atoms in propane, shown below, is

120

$$H_3C-CH_2-CH_3$$

a. 180^0 c. 120^0
b. 109.5^0 d. 100^0

6. The two most abundant elements in the Earth's crust are:

a. C & H   c. C & N
b. C & Si   d. H & Na

7. Indicate which two elements make up about 75 % of the Earth.

a. H and C   c. C and Si
b. C and Fe   d. O and Si

8. Indicate which of the following properties is associated with organic compounds. There may be more than one answer.

a. soluble in water
b. most of them will burn
c. they conduct electricity
d. bonding is almost entirely covalent

9. Indicate which of the following properties is associated with inorganic compounds. There may be more than one answer.

a. many conduct electricity
b. reactions are very fast
c. C is the major component in these compounds
d. most are solids with high melting points

COMPLETION. Write the word, phrase or number in the blank space or draw the appropriate structure in answering the question.

1. A simple aldehyde molecule contains a _____ group, with the requirement that the (carbonyl) carbon atom must be the _____ carbon atom in the molecule.

2. A simple ketone contains a _____ group and it must be positioned at an _____ position in the molecule.

3. Cholesterol is made in the liver (natural) and also has been made synthetically in the chemical laboratory. A comparison of the structure and its chemical characteristics shows that the cholesterol made by these entirely different routes is _____ .

4. Draw the condensed structural formula for a primary and a tertiary amine that has three carbons and the necessary number of hydrogens.

5. Predict the HCH, HCC and the OCC bond angles in the ketone, acetone.

121

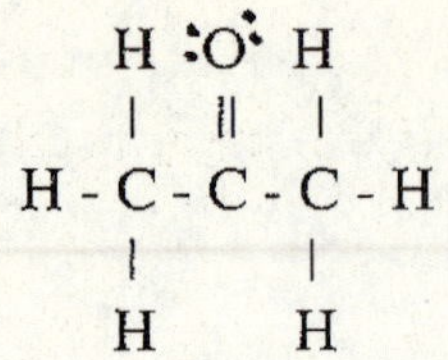

```
      H  :Ö: H
      |   ||  |
  H - C - C - C - H
      |       |
      H       H
```

<u>Bond angle</u>

a. HCH _____
b. HCC _____
c. OCC _____

6. Draw the Lewis structures for the following molecules.

    a. HCl
    b. formic acid, HCOOH
    c. H_3CCH_2CHO
    d. $H_2NCH_2CH_3$

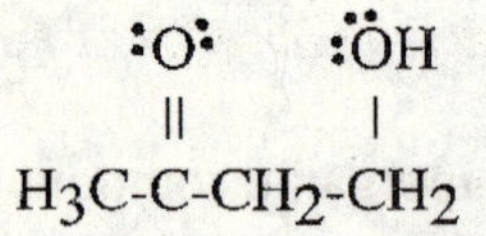

7. Identify the functional group(s) in the following molecules.

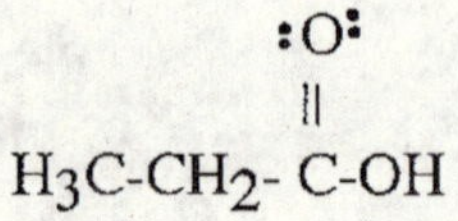

```
              :O:      :ÖH
               ||       |
a.      H3C-C-CH2-CH2
```

```
                    :O:
                     ||
b.      H3C-CH2- C-OH
```

```
c.      CH3 :O:    :O:
             |   ||     ||
        HN-CH2C-CH2-CH
```

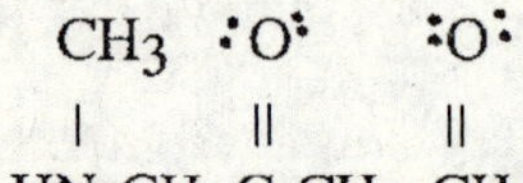

8. Draw at least four condensed structural formulas for the amines with the molecular formula of C_4NH_{11}.

9. Think about the position of C in the Periodic Table and indicate at least two features of carbon chemistry that permits it to make so many different compounds.
    i. _____
    ii. _____

10. Draw the condensed structure for a compound with the molecular formula indicated.
    a. a secondary alcohol with formula, C_3H_8O.
    b. a tertiary amine with formula, $C_4H_{11}N$.
    c. a molecule that has two aldehyde groups (a dialdehyde) with the formula, $C_4H_6O_2$.

122

        Indicate whether the statements are true or false. The number of the question refers to the section of the book that the question was taken from.

10.1 What Is Organic Chemistry?
    (a) All organic compounds contain atoms of carbon.
    (b) The majority of organic compounds are built from carbon, hydrogen, oxygen and nitrogen.
    (c) By number of atoms, carbon is the most abundant element in the Earth's crust?
    (d) Most organic compounds are soluble in water.

10.2 Where Do We Obtain Organic Compounds?
    (a) Organic compounds can only be synthesized in living organisms.
    (b) Organic compounds synthesized in the laboratory have the same chemical and physical properties as those synthesized in living organisms.
    (c) Chemists have synthesized many organic compounds that are not found in nature.

10.3 How Do We Write Structural Formulas of Organic Compounds?
    (a) In organic compounds, carbon normally has four bonds and no unshared pairs of electrons.
    (b) When found in organic compounds, nitrogen normally has three bonds and one unshared pair of electrons.
    (c) The most common bond angles about carbon in organic compounds are approximately 109.5° and 180°.

10.4 What Are Functional Groups?
    (a) A functional group is a group of atoms in an organic molecule that undergoes a predictable set of chemical reactions.
    (b) The functional group of alcohols, aldehydes, and ketones have in common the fact that each contains a single oxygen atom.
    (c) A primary alcohol has one -OH group, a secondary alcohol has two –OH groups, and a tertiary alcohol has three –OH groups.
    (d) There are two alcohols with the molecular formula C_3H_8O.
    (e) There are three amines with the molecular formula C_3H_7N.
    (f) Aldehydes, ketones, carboxylic acids, and esters all contain a carbonyl group.
    (g) A compound with the molecular formula of C_3H_6O may be either an aldehyde, a ketone, or a carboxylic acid.
    (h) Bond angles about the carbonyl carbon of an aldehydes, ketones, carboxylic acids, and esters are approximately 109.5°.
    (i) The molecular formula of the smallest aldehyde is C_3H_6O, and that of the smallest ketone is also C_3H_6O.
    (j) The molecular formula of the smallest carboxylic acid is $C_2H_4O_2$.

123

ANSWERS TO SELF-TEST QUESTIONS

MULTIPLE CHOICE

1. only an alcohol
2. b, d
3. a
4. b
5. b. It is important to understand that in simple drawings, the carbons appear in a line (180^0); however, the (CCC) angles are 109.5^0.
6. b
7. d
8. b, d
9. a, b, d

COMPLETION

1. carbonyl, terminal or end
2. carbonyl, internal
3. identical
4. primary amine $H_3CCH_2CH_2\ddot{N}H_2$

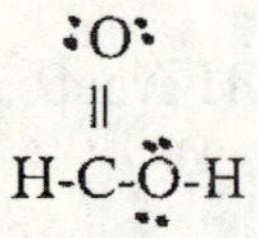

   tertiary amine $H_3C\ddot{N}CH_3$

   $\quad\quad\quad\quad\quad\quad\quad |$

   $\quad\quad\quad\quad\quad\quad CH_3$

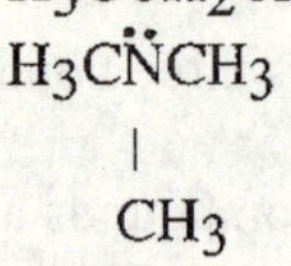

5. a. 109.5^0
   b. 109.5^0
   c. 120^0

6. a. 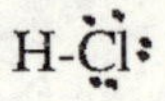   H-C̈l:

   b.
   ```
        :O:
        ‖
   H-C-Ö-H
        ··
   ```

   c.
   ```
        H H:O:
        | |  ‖
   H-C-C-C-H
        | |
        H H
   ```

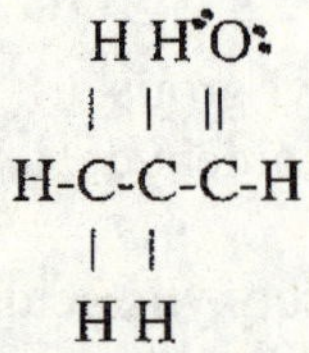

   d. 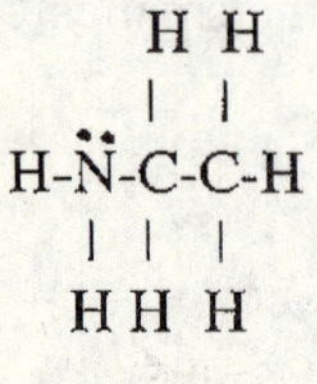
   ```
           H H
           | |
   H-N̈-C-C-H
      | | |
      H H H
   ```

7. a.   ketone and alcohol

b. carboxylic acid

c. ketone, aldehyde and amine

8.

$$
\overset{\text{H}}{\underset{\ddot{}}{\text{H}_3\text{CCH}_2\text{CH}_2\text{NCH}_3}}
$$

H₃CCH2CH2CH2N̈H2

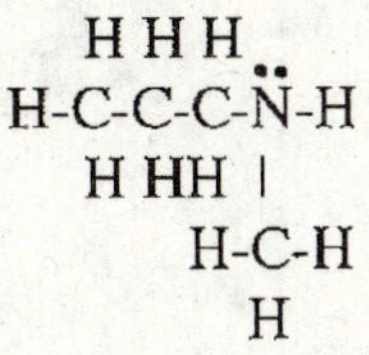

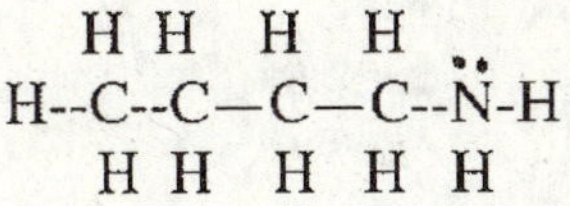

```
 H H H  ..
H-C-C-C-N-H
 H HH |
     H-C-H
      H
```

```
 H  H  H  H  ..
H--C--C—C—C--N-H
 H  H  H  H  H
```

[(CH₃)]₂N̈CH₂CH₃

```
      H
   H -C-H
      |  H H
   N--C--C--H
      |  H H
   H-C-H
      H
```

[(CH₃)]₂CHN̈H(CH₃)

```
     H    H H
   H-C--N-C-CH
     H   H| H
         HCH
          H
```

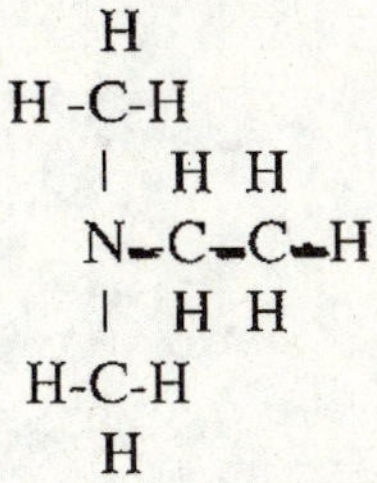

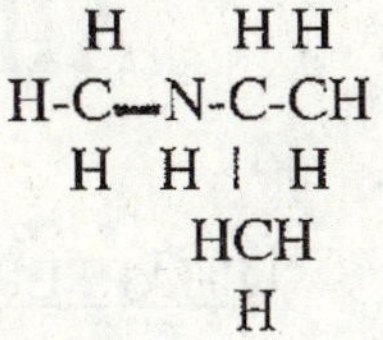

9. i. C can form stable bonds to other carbon atoms

   ii. C can form single, double and triple bonds

10.

a.
$$
\underset{\text{CH}_3\text{CHCH}_3}{\overset{:\ddot{\text{O}}\text{H}}{|}}
$$

b (CH₃)₂N̈CH₂CH₃

c. CHÖCH₂CH₂CHÖ:

QUICK QUIZ

10.1

| (a) True | (b) True | (c) False | (d) False |

10.2

| (a) False | (b) True | (c) True |

10.3

| (a) True | (b) True | (c) False |

10.4

(a) True	(b) True	(c) False	(d) True
(e) False	(f) True	(g) False	(h) False
(i) False	(j) False		

125

Gasoline - $4.00+/gal
   - Sign at gas station in
     Bowling Green, OH; Summer, 2008

11

Alkanes

CHAPTER OBJECTIVES

After you have studied the chapter and worked the assigned exercises in the text and the study guide, you should be able to do the following.

1. Name and draw the first ten unbranched alkanes, noting trends in their physical characteristics.

2. Become proficient in drawing all members of the alkane family using the line-angle formulas.

3. Consider the molecular formula for some alkanes with greater than 4 carbons and identify any constitutional (structural) isomers.

4. Using the IUPAC rules of nomenclature for saturated hydrocarbons, write the names for a variety of linear, branched and cycloalkanes.

5. Write out the structural formulas of molecules for which the IUPAC names are given.

6. Memorize the prefixes and suffixes that are used in naming molecules and provide information about the carbon-carbon bond and the functional group.

7. Become proficient in drawing the conformations for cyclic hydrocarbons and in distinguishing between axial and equatorial positions in cycloalkanes.

8. Recognize molecules that can exhibit cis- and trans-isomers.

9. Characterize the most important chemical and physical properties of alkanes.

10. Outline the most important characteristics of the (C-C) bond that is found in all alkane molecules.

11. Define the important terms and comparisons in this chapter and give specific examples where appropriate.

IMPORTANT TERMS AND COMPARISONS

Hydrocarbons
Saturated and Unsaturated Hydrocarbons
Aliphatic Hydrocarbons, C_nH_{2n+2}
IUPAC Systematic and Common Names
Molecular and Structural Formulas
    Of Compounds
Straight Chain and Branched Chain
Cycloalkanes
Chair and Envelope Conformations
Cis- and Trans- Isomers
Conformation and Configuration
Combustion or Oxidation of Alkanes
Chlorofluorocarbons (CFCs)
Petroleum
Octane Rating
Gasohol

Alkanes (-anes)
Line-Angle Formulas
Constitutional or Structural Isomers
Methylene Group ($-CH_2-$)
Alkyl Groups
Lewis Structures
Di-, Tri-, Tetra-, Penta- and Hexa-
Free Rotation and Conformations
Axial and Equatorial Bonds
Stereoisomers and Stereocenter
Physical and Chemical Properties
Haloalkanes
Natural Gas
Fractional Distillation
Engine "Knock"

SELF-TEST QUESTIONS

MULTIPLE CHOICE. In the following exercises, select the correct answer from the choices listed. In some cases, more than one choice will be correct.

1. Consider the line-angle drawing shown below. This represents the molecule:

    a.   Heptane
    b.   3-Methylhexane
    c.   2-Methylhexane
    d.   4-Methylhexane

2. Indicate which of the following molecular formulas represents a hydrocarbon.

    a. C_3H_7Br
    b. C_3H_6O
    c. C_3H_8
    d. C_3Br_8

3. Indicate which hydrocarbon has one methylene ($-CH_2$ unit) more than in $CH_3CH_2CH_2CH_2CH_2CH_2CH_3$.

a. butane   c. propylene

b. octane   d. heptane

4. Indicate which of the following is not a constitutional isomer for C5H12. Redraw these as a line-angle formula to help to distinguish them.

a. CH3CH2CH2CH2CH3

c.
$$
\begin{array}{c}
CH3 \\
| \\
CH3\text{-}C\text{-}CH3 \\
| \\
CH3
\end{array}
$$

b.
$$
\begin{array}{c}
CH3 \\
| \\
HCCH2CH3 \\
| \\
CH3
\end{array}
$$

d.
$$
\begin{array}{c}
CH3CHCH2CH2CH3 \\
| \\
CH3
\end{array}
$$

5. Which of the following structural formulas represent the same compound?

$$
\begin{array}{c}
CH2\text{-}CH2\text{-}CH2\text{-}CH3 \\
| \\
H3C\text{-}CH \\
| \\
CH3
\end{array}
$$

$$
\begin{array}{c}
CH3 \\
| \\
H3C\text{-}CH2\text{-}CH2CH2\text{-}CH \\
| \\
CH3
\end{array}
$$

$$
\begin{array}{c}
CH2\text{-}CH3 \\
| \\
CH2 \\
| \\
CH2\text{-}CH2\text{-}CH2 \\
| \\
CH3
\end{array}
$$

(i)   (ii)   (iii)

a. all are different   c. (i) and (iii)

b. (i) and (ii)   d. all are equivalent

6. The IUPAC name for the compound below is:

$$
\begin{array}{c}
H3C\text{-}CH2\text{-}CH2\text{-}CH2 \\
| \\
CHCH2CH(CH3)2 \\
| \\
CH2\text{-}CH3
\end{array}
$$

a. 2-Methyl-4-propylheptane   c. 4- Ethyl-2-methyloctane

b. 4-Ethyl-7-methylhocttane   d. Methyl-decane

7. The structure for 5-Bromo-2-chloroheptane is:

128

a. 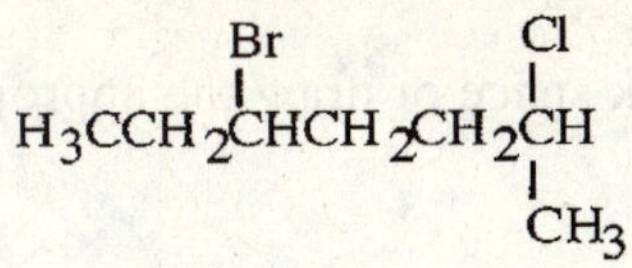

c. 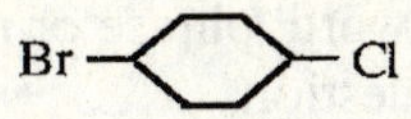

b. 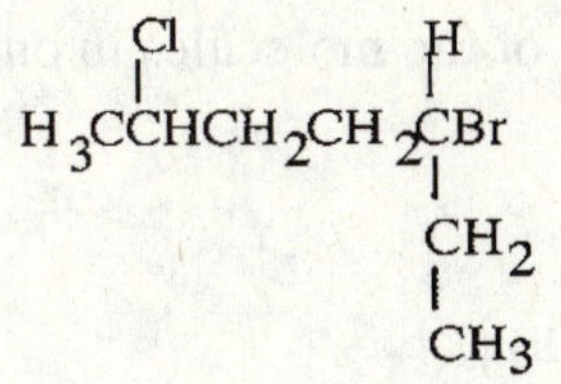

d. 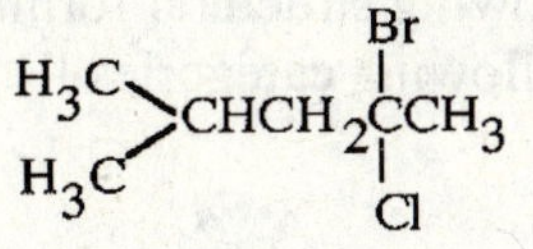

8. The combustion of pentane produces:

a. an alcohol
b. carbon dioxide and water
c. pentene
d. ethane

9. Indicate which of the following sets of molecules are constitutional isomers.

a. $CH_3CH_2OCH_3$ and $CH_3CH_2CH_2OH$
b. cyclohexane and $CH_2CHCH_2CH_2CH_2CH_3$
c. CH_3OH and CH_3CH_2OH
d. CH_3CH_2CHO and $CH_3CH_2CH_2OH$

10. Give the IUPAC name for this molecule.

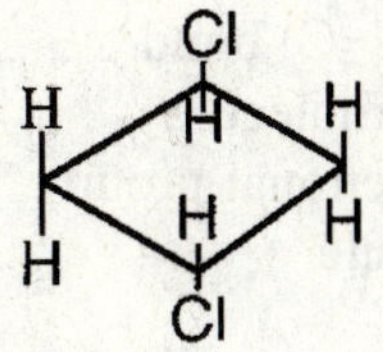

a. *trans*-1,3-Dichlorocyclobutane
b. *cis*-1,2-Dichlorocyclobutane
c. *cis*-1,3-Dichlorocyclobutane
d. Cyclobutane

11. Indicate how many bromines in this brominated sugar are in the axial positions.

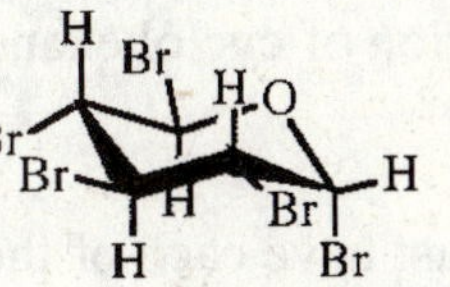

a. 1
b. 3
c. 2
d. 4
e. 5

12. Consider the bonding in octane. This involves only (C-C) and (C-H) bonds. Indicate which of the following is (are) true about the bonding?

a. All bonds are single bonds
b. There is free rotation about the bonds
c. The angle between the bonds is 109.5 degrees.
d. There is restricted rotation in the (C-C) bonds

COMPLETION. Write the word, phrase or number in the blank space or draw the appropriate structure in answering the question.

1. Examine the following structural formulas and identify each of the molecules in one or more of the following categories listed.

(i) $CH_3CH_2CH_2CH_2CH_3$

(iv)

(ii) $H_3C\ddot{O}CH_2CH_3$

(iii)

(v)

    a.   Hydrocarbon molecule   _____

    b.   Contains aldehyde functional group _____

    c.   Halogenated hydrocarbon   _____

    d.   Unsaturated (C=C) bond   _____

    e.   Straight chain molecule   _____

    f.   Contains a functional group   _____

    g.   Acyclic molecule   _____

2. The outstanding chemical characteristic of alkanes is their _____.

3. Isomers are defined as molecules which have the same _____, but different _____.

4. The _____ positions in the chair conformation of cyclohexane are found pointing up and down, respectively, on adjacent carbons.

5. A *cis*-isomer for 1,2-dimethylcyclopentane must have each of the methyl groups directed in the _____ direction on the two _____ carbon atoms at positions 1 and 2.

6. The boiling point of a series of alkanes increases as the molecular weight of the alkane _____.

7. Cyclohexane in the chair form has _____ (a number) hydrogens "pointing" up or down called _____ hydrogens, in addition to the _____ hydrogens "pointing" approximately in the "plane" of the ring called _____ hydrogens.

8. *Cis*- and *trans*-isomers are examples of _____ isomers.

130

9. Write the IUPAC name for the following molecules:

a. 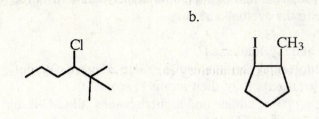   b.   c..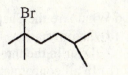

10. Draw the structures for the following compounds, showing all the atoms.

    a. 3-Chloro-2-methylpentane   _____

    b. 1-Iodo-3,3-dimethylcyclobutane _____

    c. 4-Isopropylheptane   _____

11. A saturated hydrocarbon can exist in a number of _____ because there is _____ rotation about a (C-C) single bond.

12. Balance the following equation for the combustion or oxidation of an alkane.

$$C_5H_{12} + O_2 = CO_2 + H_2O$$

QUICK QUIZ

Indicate whether the statements are true or false. The number of the question refers to the section of the book that the question was taken from.

11.1 How Do We Write Structural Formulas of Alkanes?
    (a) A hydrocarbon is composed only of the elements carbon and hydrogen.
    (b) Alkanes are saturated hydrocarbons.
    (c) The general formula of an alkane is C_nH_{2n+2}, where n is the number of carbons in the alkane.
    (d) Alkenes and alkynes are unsaturated hydrocarbons.

11.2 What Are Constitutional Isomers?
    (a) Constitutional isomers have the same molecular formulas and the same connectivity of their atoms.
    (b) There are two constitutional isomers with the molecular formula C_3H_8.
    (c) There are four constitutional isomers with the molecular formula C_4H_{10}
    (d) There are five constitutional isomers with the molecular formula C_5H_{12}.

11.3 How Do We Name Alkanes?
    (a) The parent name of an alkane is the name of the longest chain of carbon atoms in the alkane.
    (b) Propyl and isopropyl groups are constitutional isomers.
    (c) There are four alkyl groups with molecular formula C_4H_9.

11.5 What Are Cycloalkanes?
    (a) Cycloalkanes are saturated hydrocarbons.

(b) Hexane and cyclohexane are constitutional isomers.

(c) The parent name of a cycloalkane is the name of the unbranched alkane with the same number of carbon atoms as are in the cycloalkane ring.

11.6 What Are the Shapes of Alkanes and Cycloalkanes?

(a) Conformations have the same molecular formula and the same connectivity, but differ in the three-dimensional arrangement of their atoms in space.

(b) In all conformations of ethane, propane, butane and higher alkanes, all C-C-C and C-C-H bond angles are approximately 109.5°.

(c) In a cyclohexane ring, if an axial bond is above the plane of the ring on a particular carbon, axial bonds on the two adjacent carbons are below the plane of the ring.

(d) In a cyclohexane ring, if an equatorial bond is above the plane of the ring on a particular carbon, equatorial bonds on the two adjacent carbons are below the plane of the ring.

(e) The more stable chair conformation of a cyclohexane ring has more substituent groups in equatorial positions.

11.7 What Is *Cis-Trans* Isomerism in Cycloalkanes?

(a) *Cis* and *trans*-cycloalkanes have the same molecular formula and the connectivity of their atoms, but different orientation of groups in space.

(b) A *cis* isomer of a cycloalkane can be converted to its *trans* isomer by rotation about an appropriate carbon-carbon single bond.

(c) A *cis* isomer of a cycloalkane can be converted to its *trans* isomer by exchange of two groups on a stereocenter in the *cis*-cycloalkane

(d) Configuration refers to the arrangement in space of the atoms or groups of atoms at a stereocenter.

(e) *Cis*-1,4-dimethylcyclohexane and *trans*-1,4-dimethylcyclohexane are classified as conformations.

11.8 What Are the Physical Properties of Alkanes?

(a) Boiling points among alkanes with unbranched chains increase as the number of carbons in the chain increases.

(b) Alkanes that are liquid at room temperature are more dense than water.

(c) *Cis* and *trans* isomers have the same molecular formula, the same connectivity, and the same physical properties.

(d) Among alkane constitutional isomers, the least branched isomer generally has the lowest boiling point.

(e) Alkanes and cycloalkanes are insoluble in water.

(f) Liquid alkanes are soluble in each other.

11.9 What Are the Characteristic Reactions of Alkanes?

(a) Combustion of alkanes is an endothermic reaction.

(b) The products of complete combustion of an alkane are carbon dioxide and water.

(c) Halogenation of an alkane converts it to a haloalkane.

11.10 What Are Some Important Haloalkanes?

(a) The Freons are members of a class of organic compounds called chlorofluorocarbons (CFCs).

132

(b) An advantage of Freons as heat-transfer agents in refrigeration systems, propellants in aerosol sprays, and solvents for industrial cleaning is that they are nontoxic, nonflammable, odorless, and noncorrosive.

(c) Freons in the stratosphere interact with ultraviolet radiation, and thereby set up chemical reactions that lead to the destruction of the stratospheric ozone layer.

(d) Alternative names for the important laboratory and industrial solvent CH_2Cl_2 are dichloromethane, methylene chloride, and chloroform.

ANSWERS TO SELF-TEST QUESTIONS

MULTIPLE CHOICE

1.	b	7.	a (and b) are the same molecule
2.	c	8.	b
3.	b	9.	a, b
4.	d	10.	a
5.	b	11.	a
6.	c	12.	a, b, c

COMPLETION

1.
   a. i only; all other molecules contain other elements, in addition to C and H.
   b. none
   c. iii, v
   d. iii
   e. i, ii
   f. ii, iv; iii and v also, if alkyl halide is considered as a functional group
   g. i, ii, iii, v

2. lack of reactivity (except for burning or complete oxidation)
3. molecular formula, structural formula
4. axial
5. same, adjacent
6. increases
7. 6, axial, 6, equatorial
8. stereo
9.
   a. 3-Chloro-2,2-dimethylhexane
   b. *cis*-1-Iodo-2-methylcyclopentane
   c. 2-Bromo-2,5-dimethylhexane

133

10.

a. $H_3C-CH_2-CH-CHCH_3$

with substituents Cl and CH₃ on the CH and CH carbons

b.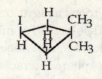

c. $H_3CH_2CH_2CCHCH_2CH_2CH_3$
     with a CH branch bearing H_3C and CH_3

11.   conformations, free
12. $1\ C_5H_{12}\ +\ 8\ O_2\ =\ 5\ CO_2\ +\ 6\ H_2O$

QUICK QUIZ

11.1
    (a) True   (b) True   (c) True   (d) True
11.2
    (a) False   (b) False   (c) False   (d) False
11.3
    (a) True   (b) True   (c) True
11.5
    (a) True   (b) False   (c) True
11.6
    (a) True   (b) True   (c) True   (d) True
    (e) True
11.7
    (a) True   (b) False   (c) True   (d) True
    (e) False
11.8
    (a) True   (b) False   (c) False   (d) False
    (e) True   (f) True
11.9
    (a) False   (b) True   (c) True
11.10
    (a) True   (b) True   (c) True   (d) False

On Chemistry:
It's hard and it's hard, ain't it
hard, Good Lord
   - Woody Guthrie
     "Hard, Ain't It Hard," 1952

12

Alkenes and Alkynes

CHAPTER OBJECTIVES

Chapters 11 and 12 begin to present some of the substance of organic chemistry. This includes not only the formal naming of, but also the drawing of structures for organic molecules containing a particular functional group. In addition, there will be an increasing emphasis on recognizing the physical properties, and especially the chemical reactions, that can take place as a direct consequence of the presence of the functional group (i.e., the "reactive center"). The strategies and learning patterns developed in studying the material in this chapter will also be especially useful in subsequent chapters.

After you have studied the chapter and worked the assigned exercises in the text and the study guide, you should be able to do the following specific objectives.

1.   Recognize the structural differences in alkenes, alkynes and distinguish them from aromatic molecules.

2.   Given the structural formulas for both alkenes and alkynes, name the unsaturated molecules using the IUPAC rules.

3.   Given the IUPAC name for an alkene or alkyne, draw the structural formula that shows all the atoms.

4.   Characterize the addition reactions in terms of the proposed intermediates and the products that occur as H_2, HCl and Cl_2 react with an alkene.

5.   Use Markovnikov's rule, as it applies to the addition of an unsymmetrical small molecule (HX, X = Cl, Br, OH) to an alkene to predict the reaction products.

6. Outline the mechanism in which ethylene and substituted ethylene undergo chain-growth polymerization to form useful polymers.

7. Define the important terms and comparisons and give specific examples where appropriate.

IMPORTANT TERMS AND COMPARISONS

Unsaturated and Saturated Hydrocarbon
Alkane, Alkene and Alkyne
Monomer and Polymer
Ethylene and Polyethylene (Structures)
Dienes, Trienes and Polyenes
IUPAC Nomenclature
Thermal Cracking (of a Saturated Hydrocarbon)
Cis-Trans Isomerization
Cycloalkenes
Carbocation (Primary, Secondary and Tertiary)
Isoprene Units and Terpenes
Low Density Polyethylene (LDPE)
Plastics

Addition Reactions For Alkenes
a. Hydrogen Halides (HCl, HBr, HI)
b. Acid-catalyzed Hydration (H_2O)
c. Halogenation (Cl2, Br2)
d. Catalytic Hydrogenation (H2)
e. Chain-Growth Polymerization
Double Bond and Restricted Rotation
Reaction Mechanism
Markovnikov's Rule
Oxonium Ion
Regioselective Reaction
High Density Polyethylene (HDPE)

Before challenging your comprehension of this material by working out the SELF-TEST, a few words are in order to help sort out the major themes presented. As this and subsequent chapters are individually addressed, it will become clear that although the general objectives and the nature of the material are similar, the strategy in learning is much the same. This is especially true in Chapters 13 through 19. Each of these chapters introduces one or more families of molecules with a characteristic functional group. **IT IS THE FUNCTIONAL GROUP THAT IS THE REACTIVE CENTER OF THE MOLECULE.** That is, this is where the reaction will take place with other molecules. The functional group is the dominant factor in defining the chemical and physical properties of a compound. Compare the chapter titles with the list of functional groups. The focus of Chapter 12 is aimed at basically three major aspects of unsaturated molecules. These molecules, the alkenes, contain one or more (C=C) double bonds. The three themes are:

1. The **naming** of unsaturated molecules using the internationally accepted International Union of Pure and Applied Chemistry (IUPAC) rules. This nomenclature systematically defines the composition and structure of the molecule.

2. The **characterization of the physical properties** of each family of molecules.

3. **Identification of the reactions** that characterize each type of (unsaturated) molecule. These same general themes will reoccur in the subsequent chapters, with simply the focus reset on a different functional group.

Possibly the most troublesome area for students is developing a strategy in identifying and characterizing the reactions associated with a family of molecules. A guide or "road map" for the process, which is both simple and useful, is the following.

136

1. Identify the functional group.
2. Tabulate the reactions associated with the functional group.
3. Characterize the reaction by its general name (addition reaction, oxidation, substitution, etc.) and then tabulate any additional specifics (catalyst or heat required, follows Markovnikov's rule, etc.).

It is very helpful if the student can write out a condensed summary of the reactions for each family of molecules. An example for alkenes follows.

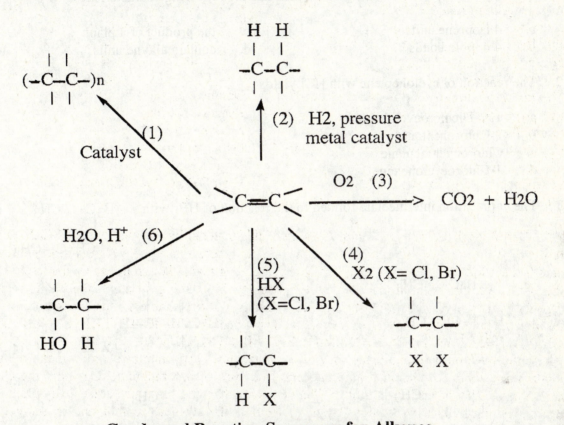

Condensed Reaction Summary for Alkenes

Reaction	Name	Comment
(1)	Polymerization	Requires appropriate catalyst; addition reaction
(2)	Catalytic hydrogenation	Addition reaction; pressure and metal catalyst
(3)	Combustion or oxidation of a hydrocarbon	CO_2 and H_2O are the common combustion products of all hydrocarbons
(4)	Bromination or chlorination	Addition reaction
(5)	Addition reaction of an HX gas	Obeys Markovnikov's rule

137

(6)	Hydration	Obeys Markovnikov's rule

SELF-TEST QUESTIONS

MULTIPLE CHOICE. In the following exercises, select the correct answer from the choices listed. In some cases, two or more answers will be correct.

1. A noncyclic terpene containing 20 carbons has the following characteristics:

   a. 4 isoprene units   c. the product of a plant
   b. 4 double bonds   d. contain alkyne units

2. The reaction of cycloheptene with HCl yields

   a. 1, 2-Dichlorocycloheptene
   b. 2-Chloroheptane
   c. Chlorocycloheptane
   d. 1-Chlorocyclohexane

3. The carbocation intermediate formed in the reaction of HBr with

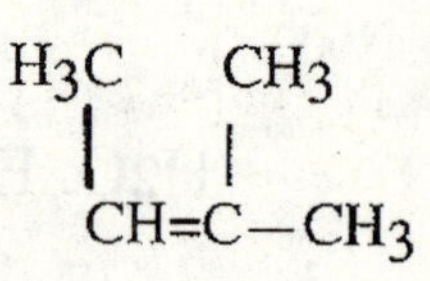

$$H_3C \quad CH_3$$
$$| \quad\quad |$$
$$CH=C-CH_3$$

is:

a. 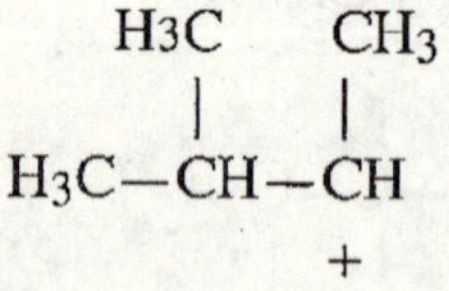
$$H_3C \quad\quad CH_3$$
$$| \quad\quad\quad |$$
$$H_3C-CH-CH$$
$$+$$

c. 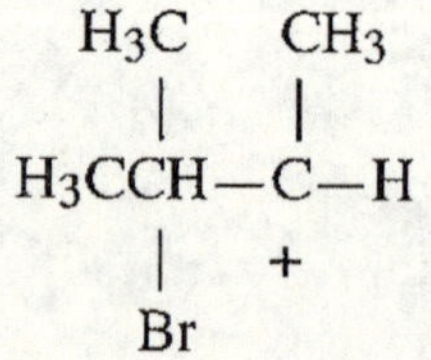
$$H_3C \quad\quad CH_3$$
$$| \quad\quad\quad |$$
$$H_3CCH-C-H$$
$$| \quad\quad +$$
$$Br$$

b. 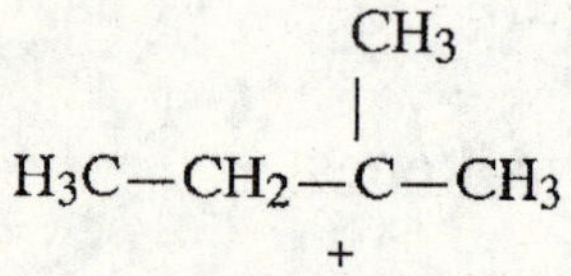
$$CH_3$$
$$|$$
$$H_3C-CH_2-C-CH_3$$
$$+$$

d. 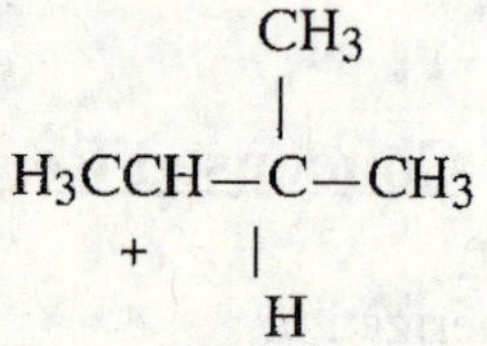
$$CH_3$$
$$|$$
$$H_3CCH-C-CH_3$$
$$+ \quad |$$
$$\quad\quad H$$

4. The principal product of the reaction of Br_2 with 1-methylcyclopentene is:

   a. 2-Bromo-1-methylcyclopentane
   b. 1, 2-Dibromo-1-methylpentane
   c. no reaction
   d. 1, 2-Dibromo-1-methylcyclopentane

5. The product of the reaction of HBr with which of the starting materials listed below yields 2-Bromo-2-methylbutane.

138

a.

CH2CHCHCH3
|
CH3

c.

CH3CHCCH3
|
CH3

b.

CH3CH2CH2CH2CH3

d.

CH3
|
CH2CCHCH2

6. The polymerization of vinyl chloride ($H_2C=CHCl$) produces:

a. unsaturated polyvinylchloride
b. saturated polyvinylchloride
c. a branched polymer
d. a linear polymer

7. An example of a secondary carbocation is:

a. CH_3^+

c.

CH3
|
H3C—C +
|
CH3

b. H3C—CH2
   +

d. H3C—CH—CH3
        +

8. The structural formula for the cis-isomer of ClCH=CCH3Cl is:

a.

c.

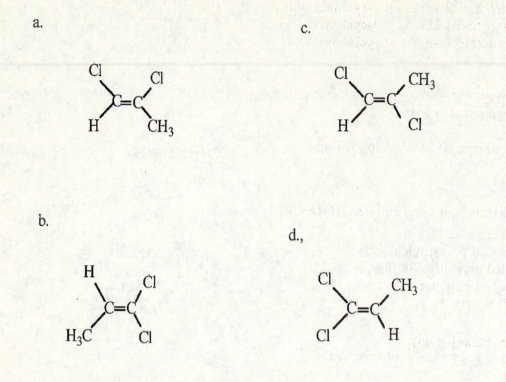

b.

d.,

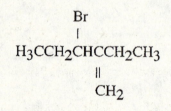

9. The reaction of HBr with which of the compounds shown below will produce H₃CCH₂CHBrCH₃:

    a. H₂C=CHCH₂CH₃   c. H₃CCH₂CH₂CH=CH₂
    b. H₃CCH=CHCH₃   d. H₃CC≡CCH₃

10. The IUPAC name for the compound below is:

$$
\begin{array}{c}
\text{Br} \\
| \\
\text{H}_3\text{CCH}_2\text{CHCCH}_2\text{CH}_3 \\
\| \\
\text{CH}_2
\end{array}
$$

    a. 3-Bromo-2-ethyl-1-pentene
    b. 3-Bromo-2-ethyl-2-hexene
    c. 3-Bromo-4-ethyl-1-pentene
    d. 3-Bromo-4-ethyl-1-hexane

11. The name of this compound is:

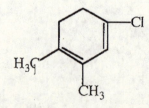

140

a.   1-Chloro-3, 4-dimethyl-2, 3-cyclohexatriene
b.   1-Chloro-3, 4-dimethyl-1, 3-cyclohexadiene
c.   3, 4-Dimethyl-1-chloro-1, 3-cyclohexadiene
d.   1-Chloroorthomethyl-1, 4-cyclohexadiene

COMPLETION. Write the word, phrase or number in the blank or draw the appropriate structure in answering the question.

1.   Give the common or IUPAC names for the following molecules.

a.

$$CH_3$$
$$|$$
$$CH_2=CCH=CH_2$$

b.   (-CH2CH2)n

c.   C5H8 (a cyclic compound)

d.

e.

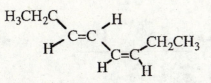

2.   Fill in the Table below with the characteristics for alkanes, alkenes and alkynes.

Characteristic	Alkanes	Alkenes	Alkynes
Two Letter Suffix to Describe Hydrocarbon Type			
Bonding Angle (degrees)			
Relative (C-C) Bond Length			
Addition Reactions (yes or no)			

141

3. Draw the structures for the products of reactions b-d. In reaction a, only indicate the products.

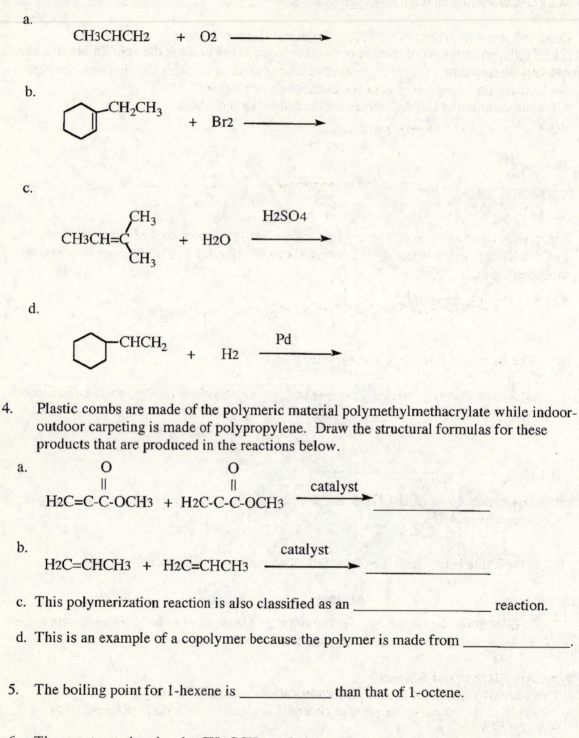

a.

   CH3CHCH2   +   O2   ⟶

b.

   $\begin{array}{c}\text{cyclohexene}\end{array}$—CH₂CH₃   +   Br2   ⟶

c.

   CH3CH=C(CH₃)(CH₃)   +   H2O $\xrightarrow{\text{H2SO4}}$

d.

   cyclohexyl—CHCH₂   +   H2 $\xrightarrow{\text{Pd}}$

4. Plastic combs are made of the polymeric material polymethylmethacrylate while indoor-outdoor carpeting is made of polypropylene. Draw the structural formulas for these products that are produced in the reactions below.

   a.

$$H_2C=\overset{\displaystyle O}{\overset{\displaystyle \|}{C}}-C-OCH_3 \ + \ H_2C-\overset{\displaystyle O}{\overset{\displaystyle \|}{C}}-C-OCH_3 \ \xrightarrow{\text{catalyst}} \underline{\hspace{3cm}}$$

   b.

   H2C=CHCH3   +   H2C=CHCH3 $\xrightarrow{\text{catalyst}}$ \underline{\hspace{3cm}}

   c. This polymerization reaction is also classified as an \underline{\hspace{3cm}} reaction.

   d. This is an example of a copolymer because the polymer is made from \underline{\hspace{3cm}}.

5. The boiling point for 1-hexene is \underline{\hspace{3cm}} than that of 1-octene.

6. The unsaturated molecule, CH3CCH, methylacetylene, can be hydrogenated, similar to that discussed for ethylene. The structure for the product of this hydrogenation is

_____. This product can then be reacted with HCl to produce a saturated alkylchloride. The structure for this final product is _____ .

7. Markovnikov's rule states _____.

8. Draw the structural formula for 1, 2, 3-Tribromo-4-chlorobutadiene.

9. Alkynes undergo the very same types of reactions as alkenes. With this in mind, draw the products for the sequence of the two reactions shown below:

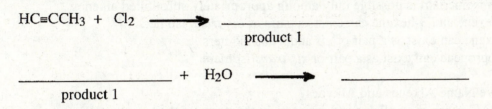

product 1

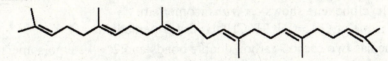

product 1

10. Squalene is a terpene compound involved in the biosynthesis of cholesterol in the body. The condensed structure is shown below and can be regarded as a polymer of isoprene monomer units.

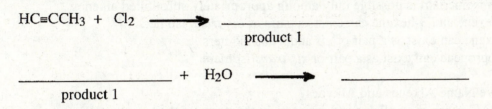

   a.   The structural formula for an isoprene unit is _____.

   b.   Although the many structures of terpenes are very diverse, they all share two common characteristics.
        (i) _____

        (ii) _____

   c.   Determine the number of isoprene units in squalene and underline these units in the structure shown.

        number of isoprene units = _____

QUICK QUIZ

Indicate whether the statements are true or false. The number of the question refers to the section of the book that the questions were taken from.

12.1 What Are Alkenes and Alkynes?
   (a) There are two classes of unsaturated hydrocarbons, alkenes and alkynes.
   (b) The bulk of the ethylene used by the chemical industry worldwide is obtained from renewable resources.
   (c) Ethylene and acetylene are constitutional isomers.
   (d) Cyclohexane and 1-hexene are constitutional isomers.

12.2 What Are the Structures of Alkenes and Alkynes?.
   (a) Both ethylene and ethane are planar molecules.
   (b) An alkene in which each carbon of the double bond has two different groups bonded to it will show *cis-trans* isomerism.
   (c) *Cis-trans* isomers have the same molecular formula but a different connectivity of their atoms.
   (d) *Cis*-2-butene and *trans*-2-butene can be interconverted by rotation about the carbon-carbon double bond.
   (e) *Cis-trans* isomerism is possible only among appropriately substituted alkenes.
   (f) Both 2-hexene and 3-hexene can exist as pairs of *cis-trans* isomers.
   (g) Cyclohexene can exist as a pair of *cis* and *trans* isomers.
   (h) 1-Chloropropene can exist as a pair of *cis-trans* isomers.

12.3 How Do We Name Alkenes and Alkynes?
   (a) The IUPAC name of an alkene is derived from the name of the longest carbon chain that contains the carbon-carbon double bond.
   (b) The IUPAC name of $CH_3CH=CHCH_3$ is 1-methylpropene.
   (c) 2-Methyl-2-butene shows *cis-trans* isomerism.
   (d) 1,2-dimethylcyclohexene shows *cis-trans* isomerism.
   (e) The IUPAC name of $CH_2=CHCH=CHCH_3$ is 1,3-pentadiene.
   (f) 1.3-Butadiene has two carbon-carbon double bonds and $2^2 = 4$ stereoisomers are possible for it.

12.4 What Are the Physical Properties of Alkenes and Alkynes?.
   (a) Alkenes and alkynes are nonpolar molecules.
   (b) The physical properties of alkenes are similar to those of alkanes of the same carbon skeletons.
   (c) Alkenes that are liquid at room temperature are insoluble in water and when added to water, will float on water.

12.5 What Are Terpenes?
   (a) Terpenes are identified by their carbon skeleton, namely one that can be divided into five-carbon units, each identical with the five-carbon skeleton of isoprene.
   (b) Isoprene is a common name for 2-methyl-1,3-butadiene.
   (c) Both geraniol and menthol (Figure 12.1) show cis-*trans* isomerism.
   (d) *Cis-trans* isomerism is not possible in myrcene.

12.6 What Are the Characteristic Reactions of Alkenes?.
   (a) Complete combustion of an alkene gives carbon dioxide and water.
   (b) Addition reactions of alkenes involve breaking one of the bonds of the carbon-carbon double bond and formation of two new single bonds.
   (c) Markovnikov's rule refers to the regioselectivity of addition reactions of carbon-carbon double bonds.
   (d) According to Markovnikov's rule, in the addition of HCl, HBr, or HI to an alkene, hydrogen adds to the carbon of the double bond that already has the greater number of hydrogen atoms bonded to it, and the halogen adds to the carbon that has the lesser number of hydrogen atoms bonded to it.

144

(e) A carbocation is carbon with four bonds to it and that bears a positive charge.
(f) The carbocation derived from ethylene is $CH_3CH_2^+$.
(g) The reaction mechanism for the addition of a halogen acid to an alkene is divided into two steps, (1) formation of a carbocation and (2) reaction of the carbocation with halide ion, which complete the reaction.
(h) Acid-catalyzed addition of H_2O to an alkene is called *hydration*.
(i) If a compound fails to react with Br_2, it is unlikely that the compound contains a carbon-carbon double bond.
(j) Addition of H_2 to a double bond is a reduction reaction.
(k) Catalytic reduction of cyclohexene gives hexane.
(l) According to the mechanism presented in the text for acid-catalyzed hydration of an alkene, the H and –OH groups added to the carbon-carbon double bond both arise from the same molecule of H_2O.
(m) The conversion of ethylene, $CH_2=CH_2$, to ethanol, CH_3CH_2OH is an oxidation reaction.
(n) Acid-catalyzed hydration of 1-butene gives 1-butanol. Acid-catalyzed hydration of 2-butene gives 2-butanol.

12.7 What Are the Important Polymerizations Reactions of Ethylene and Substituted Ethylenes?
(a) Ethylene contains one carbon-carbon double bond and polyethylene contains many carbon-carbon double bonds.
(b) All C-C-C bond angles in both LDPE and HDPE are approximately 120°.
(c) Low-density polyethylene (LDPE) is a highly branched polymer.
(d) High-density polyethylene (HDPE) consists of carbon chains with little branching.
(e) The density of polyethylene polymers is directly related to the degree of chain branching; the greater the branching, the lower the density of the polymer.
(f) PS and PVC are currently recycled.

ANSWERS TO SELF-TEST QUESTIONS

MULTIPLE CHOICE

1.	a, b, c		7.	d
2.	c		8.	a
3.	b		9.	a, b
4.	d		10.	a
5.	c		11.	b
6.	b, d		12.	b, most stable; a, least stable

COMPLETION

1.
   a. Isoprene or 2-Methyl-1, 3-butadiene
   b. Polyethylene
   c. Cyclopentene
   d. 1, 4-Cyclohexadiene
   e. trans, cis-3, 5-Octadiene

145

2.

Characteristic	Alkanes	Alkene	Alkynes
Suffix to Describe Class Of Hydrocarbon	-an-	-en-	-yn-
Bond Angle (Degrees)	109.5	120	180
Relative (C-C) Bond Length	longest	intermediate	shortest
Addition Reactions	no	yes	yes

3.

a. $CH3CHCH2 + O2 \longrightarrow CO2 + H2O$

b.

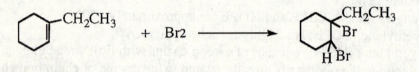

c.

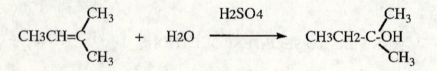

d.

4. The answers in parts (a) and (b) show the repeating unit in each polymer.

a.
```
      O        O
      ||       ||
   H3COC    COCH3
      |        |
-(-CH2-C-CH2-CH-)n-
      |
     CH3
```

b.
```
          CH3
          |
   -(-CH2-CH-)n
```

c. addition

d. two different monomer units

5. lower

6.
$$CH_3\ H \qquad\qquad CH_3\ H$$
$$|\quad\ |\qquad\qquad\quad |\quad\ |$$
$$C = C\ ;\qquad H\text{-}C\text{---}C\text{-}H$$
$$|\quad\ |\qquad\qquad\quad |\quad\ |$$
$$H\quad H\qquad\qquad\ \ Cl\quad H$$

7. When an unsymmetrical reagent such as HX adds to an alkene or alkyne, the hydrogen from HX adds to the carbon atom of the double bond which already has the greater number of hydrogen atoms.

8.
$$Cl\ \ Br\ \ Br\ \ Br$$
$$|\quad\ |\quad\ \ |\quad\ |$$
$$HC = C\text{ - }C = CH$$

9.   Product 1   Product 2

$$Cl\quad Cl \qquad\qquad\qquad\qquad Cl\quad Cl$$
$$|\quad\ |\qquad\qquad\qquad\qquad\quad |\quad\ |$$
$$H\text{-}C = C\text{-}CH_3 \qquad\qquad\qquad H\text{-}C\text{ - }C\text{ -}CH_3$$
$$\qquad\qquad\qquad\qquad\qquad\qquad\qquad |\quad\ |$$
$$\qquad\qquad\qquad\qquad\qquad\qquad\quad H\quad OH$$

10. CH_3
                          |
    a $H_2C{=}C\text{-}C{=}CH_2$
                          |
                          H

b. (i) They are synthesized in plants and are therefore called "natural products".
   (ii) All follow the isoprene rule.

c. 6 isoprene units

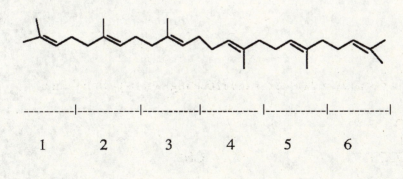

QUICK QUIZ

12.1

(a) False	(b) False	(c) False	(d) True

12.2

(a) False	(b) True	(c) False	(d) True
(e) False	(f) True	(g) False	(h) True

12.3

(a) True	(b) False	(c) False	(d) False
(e) True	(f) False		

12.4

(a) True	(b) True	(c) True

12.5

(a) True	(b) True	(c) True	(d) True

12.6

(a) True	(b) True	(c) True	(d) True
(e) False	(f) True	(g) True	(h) True
(i) True	(j) True	(k) False	(l) False
(m) False	(n) False		

12.7

(a) False	(b) False	(c) True	(d) True
(e) True	(f) True		

A great teacher is one who makes
himself progressively unnecessary.
  - Thomas Carruthers

13

Benzene and Its Derivatives

CHAPTER OBJECTIVES

After you have studied the chapter and worked the assigned exercises in the text and the study guide, you should be able to do the following.

1. Recognize aromatic compounds and aromatic groups in larger, more extended structures.

2. Draw the Lewis structure of the aromatic molecule, benzene, in its alternative Lewis contributing structures and in its resonance hybrid form.

3, Given the structural formulas for a variety of aromatic molecules, name the molecules using the IUPAC rules of nomenclature.

4. Given the IUPAC name for aromatic compounds, draw their structures, showing all atoms.

5. Characterize the type of chemistry that aromatic molecules exhibit, including substitution reactions involving halogenation, nitration and sulfonation.

6. Write out the acid-base reaction for a phenol titrated with NaOH.

7. Contrast the chemical reactions associated with alkenes with those for aromatic molecules and indicate their products.

8. Define the important terms and comparisons in this chapter and give specific examples where appropriate.

IMPORTANT TERMS AND COMPARISONS

Aromatic or Arene Compounds

Resonance, Resonance Hybrid and Aromatic Sextets

Ortho (o), Meta (m) and Para (p)

Polynuclear Aromatic Hydrocarbons (PAHs)

DDT (Dichlorodiphenyltrichloroethane)

Addition Versus Substitution Reactions

Phenol and the Phenyl Group

Ka and pKa Values and Acidity

Kekule Structure

Aryl, Phenyl, and Benzyl Groups

Carcinogen, Diolepoxide and Smoking

TNT (2, 4, 6-Trinitrotoluene and
    Trinitroglycerin (Nitroglycerin)

Antioxidants, Vitamin E, BHT and BHA

Acid-Base Chemistry of Phenol

Autooxidation - Radical Chain Reaction

Reactions For Aromatic Molecules
    Halogenation
    Nitration
    Sulfonation
    Radical, Chain Initiation and Propagation

SELF-TEST QUESTIONS

MULTIPLE CHOICE.   In the following exercises, select the correct answer from the choices listed. In some cases, two or more answers may be correct.

1.   Indicate which of the following properties is a characteristic of benzene and its derivatives.

    a.   Soluble in organic solvents and insoluble in water
    b.   Burns in air to form carbon dioxide and water
    c.   Bonding is essentially covalent
    d.   Reacts with acids and bases

2.   The nitration, halogenation and sulfonation reaction of benzene is generally called a

    a   substitution reaction   c.   elimination reaction
    b.   addition reaction   d.   dehydration reaction

3.   The name of the following molecule is;

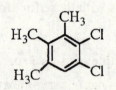

    a.   2, 3-Dichloro-5, 6-dimethyltoluene
    b.   5, 6-Dichloro-2, 3-dimethyltoluene
    c.   2, 3-Dichloro-4, 5, 6-trimethyltoluene
    d.   1, 2, 3-Trimethyl-5, 6-dichlorobenzene

150

4. The individual (C-C) bonds in each Lewis structure for benzene are either single or double bonds. The (C-C) bond length in benzene is actually

    a, alternating single and double bonds
    b. a bond type midway between a single and double bond
    c. primarily single bonds
    d. primarily double bonds

5. In polynuclear aromatic hydrocarbons (PAHs), the aromatic rings

    a. share a single carbon atom.
    b. always share three or more carbon atoms.
    c. are held together by a single linking carbon atom.
    d. share one or more sides of the aromatic rings.

6. Reaction of I_2 with the aromatic ring in the protein, thyroglobulin, starts the synthesis of thyroxine. Indicate what type of reaction this is.

    a. addition reaction   c. substitution
    b. elimination   d. oxidation

7. In a disubstituted benzene, the greatest possibility for the substituted groups "bumping into" each other is when the two groups are in

    a. ortho positions   c. meta positions
    b. para positions   d. either ortho, meta or para

8. Phenol exhibit a significant acidic character, much more than found in alcohols. The pKa for phenol is 9.95. At pH 9.95, the relative ratio of phenol/phenoxide is:

    a. 50/50   c. 100% phenol
    b. 30/70   d. 100% phenoxide

9. Indicate which of the following reactions involves the participation of a radical.

    a. autoxidation of a polyunsaturated fatty acid
    b. acid-base reaction of phenol
    c. nitration of benzene
    d. halogenation of benzene

COMPLETION. Write the word, phrase or number in the blank or draw the appropriate structure in answering the question.

1. Draw the structures for these common aromatic units.

    a. phenol   _____

151

b.   aniline   _____

c   benzaldehyde   _____

d.   benzoic acid   _____

e.   anisole   _____

f.   toluene   _____

2.   Draw the structures for these common groups or substituents.

    a.   phenyl   _____

    b.   benzyl   _____

    c.   phenoxide   _____

3.   Name the three different isomers that are possible for a benzene ring that has one nitro and one chloro substituent in place of a hydrogen. Indicate the site numbers for these arrangements.

    Names   Numbers for the substitution sites

a   _____   _____

b.   _____   _____

c.   _____   _____

4.   Write out the complete equations for these reactions.

    a.   chlorination of benzene

    b.   sulfonation of benzene, followed by treatment with NaOH

    c.   nitration of benzene

    d.   titration of phenol with NaOH

    e.   conversion of nitrobenzene to aniline

5.   Draw the following molecules.

    a.   Dibenzylmethane

    b.   4-Bromo-2-chlorobenzaldehyde

152

c. p-Methylnitrobenzene (or p-Nitrotoluene)

d. 6-Bromo-2, 3-Dichlorotoluene

6. Foods can undergo a reaction called _____ if they have carbon-carbon double bonds and are exposed to other reagents and _____ gas. These reactions are initiated by either _____ or light, which converts the unsaturated molecule to a _____. The unpaired electron resides in the molecule on the _____ atom. This leads to an _____ _____ reaction, which can lead to spoilage. A common defense to inhibit this reaction from occurring in commercial foods is to add a chemical that contains a _____ functional group such as (use abbreviations) _____ or _____. Nature's "scavenger" is vitamin ___.

QUICK QUIZ
Indicate whether the statements are true or false. The number of the question refers to the section of the book that the question was taken from

13.1 What Is the Structure of Benzene?
(a) Alkenes, alkynes and arenes are unsaturated hydrocarbons.
(b) Aromatic compounds were so named because many of them have pleasant odors.
(c) According to the resonance model of bonding, benzene is best described as a hybrid of two equivalent contributing structures.
(d) Benzene is a planar molecule.

13.2 How Do We Name Aromatic Compounds?
(a) A phenyl group has the molecular formula C_6H_5, and is represented by the symbol Ph-.
(b) Para substituents occupy adjacent carbons on a benzene ring.
(c) 4-Bromobenzoic acid can be separated into *cis* and *trans* isomers.
(d) Naphthalene is a planar molecule.
(e) Benzene, naphthalene, and anthracene are polynuclear aromatic hydrocarbons (PAHs).
(f) Benzo[a]pyrene causes cancer by binding to DNA and producing a cancer-causing mutation.

13.3 What Are the Characteristic Reactions of Benzene and Its Derivatives?
(a) Benzene does not undergo the addition reactions that are characteristic of alkenes.
(b) A defining feature of aromatic compounds is that they are highly unsaturated but do not undergo characteristic alkene addition reactions.
(c) Nitration of benzene adds a $-NO_2$ group to one of the carbons of the aromatic ring.
(d) Halogenation of an alkene is an addition reaction; halogenation of an arene is a substitution reaction.

13.4 What Are Phenols?

(a) Phenols and alcohols have in common, the presence of an –OH group.

(b) Phenols are weak acids and react with strong bases to give water-soluble salts.

(c) The pK_a of phenol is smaller than that of acetic acid.

(d) Autooxidation converts an R-H group to an R-OH (hydroxyl) group.

(e) A carbon radical has only seven electrons in the valence shell of one of its carbons and this carbon bears a positive charge.

(f) A characteristic of a chain initiation step is conversion of a nonradical to a radical.

(g) Autooxidation is a radical chain process.

(h) A characteristic of chain propagation step is reaction of a radical and a molecule to form a new radical and a new molecule.

(i) Vitamin E and other natural antioxidants function by interrupting the cycle of chain propagation steps that occurs in autoxidation.

ANSWERS TO SELF-TEST QUESTIONS

MULTIPLE CHOICE

1. a, b, c (phenol is a weak acid, so add d for this special case)
2. a
3. a (the parent is toluene)
4. b

5. d
6. c
7. a
8. a
9. a

COMPLETION

1.

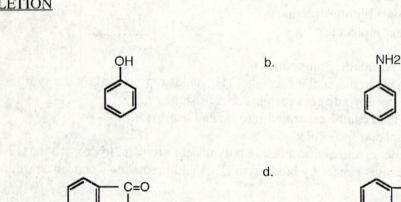

a. OH

b. NH2

c. C=O H

d. C=O OH

e. OCH3

f. CH3

2.

a.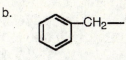

b.

c.

3.

a. o-Chloronitrobenzene   sites 1 & 2

b. m-Chloronitrobenzene   sites 1 & 3

c. p-Chloronitrobenzene   sites 1 & 4

4.

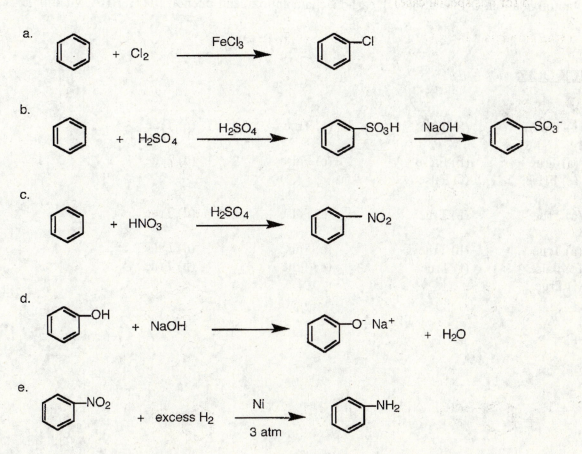

5.

a.

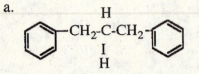

c.

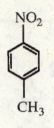

d.

b.

6.   autooxidation, O₂, heat, radical, carbon, chain antioxidant, phenol, BHT, BHA, vitamin E

QUICK QUIZ

13.1
   (a) True   (b) True   (c) True   (d) True
13.2
   (a) True   (b) False   (c) False   (d) True
   (e) False   (f) False
13.3
   (a) True   (b) True   (c) False   (d) True
13.4
   (a) True   (b) True   (c) False   (d) False
   (e) False   (f) True   (g) True   (h) True
   (i) True

More wine, less wits, you know; wine
makes a man sing even if he is a rare
scholar, makes him titter and chuckle,
aye, makes him dance a jig, and makes
him blurt out what were better kept to
himself.
 - Odyssey, XIV, 464

14

Alcohols, Ethers, and Thiols

CHAPTER OBJECTIVES

After you have studied the chapter and worked the assigned exercises in the text and the study guide, you should be able to do the following.

1. Identify compounds that contain an alcohol (R-OH), ether (R-O-R') or thiol (R-SH)functional group.

2. Given the IUPAC names for simple alcohols, ethers and thiols, draw the structural formula for each that includes all the atoms.

3. Given the structural formula for an alcohol, ether or thiol, name the compound according to the IUPAC rules.

4. Write out the reactions for the dehydration of alcohols and point out the predominant product in cases in which there are isomeric alkene products.

5. Draw the structures for primary, secondary and tertiary alcohols.

6. Write out the reactions for the oxidation of primary, secondary and tertiary alcohols and characterize the possible products formed.

7. Review the concept of hydrogen bonding and discuss the role of hydrogen bonding in understanding the boiling points of alcohols relative to hydrocarbons and solubility characteristics of alcohols and ethers in water.

8. Characterize the similarities and differences in the physical properties of alcohols, ethers, thiols and saturated hydrocarbons of comparable molecular weight.

9. Draw the structural formulas for thiols and disulfides and write out the oxidation/reduction reaction associated with them.

10. Prepare a summary sheet for the reactions of alcohols, ethers, thiols and disulfides.

11. Define the important terms and comparisons in this chapter and give specific examples where appropriate.

IMPORTANT TERMS AND COMPARISONS

Functional Group	Alcohol, Ether and Thiol
Primary, Secondary and Tertiary Alcohols	IUPAC Nomenclature
Diol, Glycols and Triol	Glycerol or Glycerin
Alfred Nobel	Nitroglycerin and Dynamite
Dehydration Reaction	Elimination Reaction
Oxidation Reaction	Hydrogen Bonding
London Dispersion Forces	Hydrophobic
Methanol, Ethanol and Isopropyl Alcohol	Aldehyde and Ketones
Carboxylic Acid	Thiols, Thioethers and Disulfides
Hydroxyl- and Sulfhydryl- Groups	Disulfides
Anesthetic	

SELF-TEST QUESTIONS

MULTIPLE CHOICE. Select the correct answer or answers from the choices listed. In some cases, more than one answer is correct.

1. The chemical oxidation of $H_3CCH_2\overset{\overset{\displaystyle OH}{|}}{C}HCH_3$ produces:

a. $CH_3CH_2\overset{\overset{\displaystyle O}{||}}{C}CH_3$

c. $CH_3CH_2\overset{\overset{\displaystyle O}{||}}{C}H$

b. $CH_3CH_2OCH_2CH_3$

d. $CO_2 + H_2O$

2. The compound that is most capable of exhibiting hydrogen bonding is:

a. $H_3COCH_2CH_3$

c. butane

158

b. CH3CH2CH2CH2OH   d. CH3CH2S-SCH3

3. Indicate the type of chemical change that occurs in the alcohol during the breath test that
    law enforcement officers use to determine alcohol levels in drivers.

    a. oxidation   c. reduction
    b. dehydration   d. substitution

4. Indicate two characteristics that make ethers potentially dangerous.

    a. solubility in water
    b. flammability
    c. potential for explosion
    d. toxicity

5. The compounds that are widely recognized for their unpleasant odors are:

    a. thiols   c. ethers
    b. alcohols   d. alkanes

6. Tertiary alcohols are resistant to oxidation because they:

    a. are already oxidized
    b. do not have a H on the C that is bonded to the alcohol group
    c. are unsaturated
    d. readily form an alkene

7. The complete oxidation of a primary alcohol produces:

    a. ketone   c. carboxylic acid
    b. aldehyde   d. cyclic ether

8. The oxidation of CH3CH2SH results in a:

    a. thiol   c. cyclic thiol
    b. disulfide (bond formation)   d. useful anesthetic

9. The order of the compounds that increasingly take part in hydrogen bonding with water is:

    a. alkane < alcohol < ether < aldehyde   c. alkane < ether < aldehyde < alcohol

    b. alcohol < alkane < aldehyde < ether   d. ether < aldehyde < alcohol < alkane

10. Oxidation of the following alcohol by potassium dichromate in sulfuric acid
    ($K_2Cr_2O_7/H_2SO_4$) yields:

159

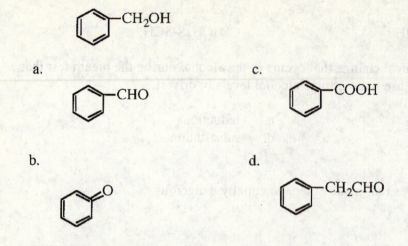

a.

b.

c.

d.

11. The oxidation of ethanol produces acetaldehyde (CH3CHO). Further oxidation of acetaldehyde yields:

   a. CHOOH
   b. CH3OH
   c. CH3COOH
   d. CH3COCH3

12. Considering the little potential for hydrogen bonding, predict what the relative boiling points for propanethiol and ethylmethylsulfide would be.

   a. comparable
   b. ethylmethylsulfide is higher
   c. propanethiol is higher
   d. cannot predict

13. Order the following compounds in terms of decreasing acidic character.

   a. alcohol > thiol > alkane > water
   b. thiol > alcohol ~ water > alkane
   c. water > alcohol > alkane > thiol
   d. thiol > alkane > water > alcohol

14. Indicate whether the following reactions will take place. For those that occur, indicate the products of the reaction.
   a. H3CSH + H⁺   =
   b. H3CSH + OH⁻ =
   c. H3CSH + Hg⁺² =
   d. H3CSH + O2   =

COMPLETION. Write the word, phrase or number in the blank or draw the appropriate structure in answering the question.

1. In the dehydration of an alcohol, two groups are removed from adjacent carbons to produce a double bond. This reaction often is referred to as an _____ reaction.

160

2. Tertiary alcohols cannot be _____ , while the oxidation of a primary alcohol yields either an _____ or a _____.

3. The relative order of boiling points for a simple alcohol and an ether, both having comparable
molecular weights is:

_____ > _____

4. The relative solubility of the propanol, propane, 1,2-propanediol and ethylmethylether in water is:

_____ > _____ > _____ > _____

5. Complete the equations by writing the appropriate structural formula.

a.
$$H_3C \quad CH_2OH$$
$$| \quad |$$
$$H_3CCHCHCH_3 \quad \xrightarrow{H_2SO_4} \quad \underline{\hspace{4cm}}$$
initial product

_____ $\xrightarrow{[O]}$ _____
initial product   final product

b. _____ $\xrightarrow[180^0 \text{ C}]{H_2SO_4}$ $H_3CCH_2\text{-}O\text{-}CH_2CH_3$

c. _____ $\xrightarrow[\text{[H]}]{\text{reduction}}$
$$\begin{array}{c} CH_3 \\ | \\ HSCH_2CHCH_2SH \end{array}$$

d.
$$\begin{array}{c} CH_2CH_3 \\ | \\ CH_3C\text{-}OH \\ | \\ CH_3 \end{array} \quad \xrightarrow[\text{heat}]{H_2SO_4}$$

e.
$$\begin{array}{c} CH_2CH_3 \\ | \\ CH_3CH\text{-}OH \end{array} \quad \xrightarrow[H_2SO_4]{K_2Cr_2O_7}$$

f. $CH_3CH_2OH \quad \xrightarrow[180^0]{H_2SO_4}$

161

g.   CH3CH2-O-CH2CH3 $\xrightarrow[\text{heat}]{\text{H2SO4}}$

h.   CH3CH2-SH $\xrightarrow{\text{[O]}}$

i. $\xrightarrow[\text{heat}]{\text{K2Cr2O7}}$ $\overset{\displaystyle O}{\overset{\displaystyle \|}{CH3CH2CH2CCH3}}$

j.   CH2--CH--CH$_2$ + 3 HNO$_3$ $\longrightarrow$
            | | |
          OH OH OH

6.   Characterize each of the following molecules as alcohols, ethers, both or neither of these.

    a. H3CCHO   _____

    b. H3CCH2CHCH2   _____

    c. HOCH2CH2OH   _____

    d.   _____

    e.   _____

    f.   _____

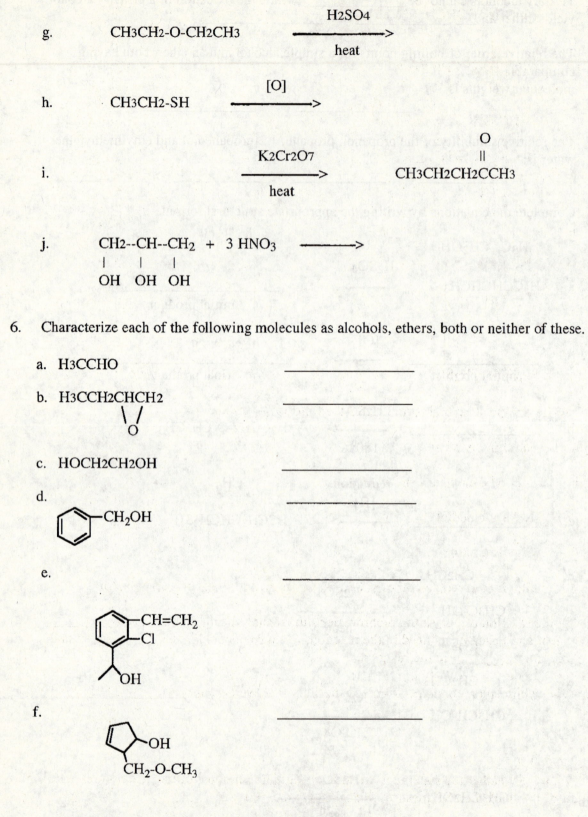

7.   Give suitable names for the following compounds.

a.   HOCH2CHOHCH2CH2OH

b.   CH3CH2-O-CH(CH3)2

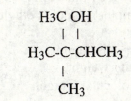

c.   H3C-C-CHCH3

d.

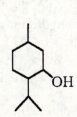

e.   H3CCH2CH2SH

8.   Write out the structural formulas for the following compounds:

   a.   Butylmethylether

   b.   4-Ethyl-3-hepten-1-ol

   c.   1, 2-Pentanediol

   d.   Ethylene oxide

   e.   Ethylene glycol, a common liquid used as an antifreeze in a car radiator.

9.   In the reaction of potassium dichromate with ethanol, the dichromate is an _____ agent and becomes reduced in the reaction, while the alcohol is the _____ agent and becomes _____ in the reaction.

10.   The cyclic ether, ethylene oxide, is especially reactive because _____.

QUICK QUIZ
      Indicate whether the statements are true or false. The number of the question refers to the section of the book that the question was taken from.

163

14.1 What Are the Structures, Names, and Properties of Alcohols?
   (a) The functional group of an alcohol is the –OH (hydroxyl) group.
   (b) The parent name of an alcohol is the name of the longest carbon chain that contains the –OH group.
   (c) A primary alcohol contains one –OH group, and a tertiary alcohol contains three –OH groups.
   (d) In the IUPAC system, the presence of three –OH groups is shown by the ending –*triol*.
   (e) A glycol is a compound that contains two –OH groups. The simplest glycol is ethylene glycol, $HOCH_2CH_2$-OH
   (f) Because of the presence of an –OH group, all alcohols are polar compounds.
   (g) The boiling points of alcohols increase with increasing molecular weight.
   (h) The solubility of alcohols in water increases with increasing molecular weight.

14.2 What Are the Characteristic Reactions of Alcohols?
   (a) The two most important reactions of alcohols are their acid-catalyzed dehydration to give alkenes, and their oxidation to aldehydes, ketones, and carboxylic acids.
   (b) The acidity of alcohols is comparable to that of water.
   (c) Water-insoluble alcohols, and water-insoluble phenols react with strong bases to give water-soluble salts.
   (d) Acid-catalyzed dehydration of cyclohexanol gives cyclohexane.
   (e) When the acid-catalyzed dehydration of an alcohol can yield isomeric alkenes, the alkene with the greater number of hydrogens on the carbons of the double bond generally predominates.
   (f) The acid-catalyzed dehydration of 2-butanol yields 1-butene as a product.
   (g) The oxidation of a primary alcohol gives either an aldehyde or carboxylic acid depending on experimental conditions.
   (h) The oxidation of a secondary alcohol gives a carboxylic acid.
   (i) Acetic acid, CH_3COOH, can be prepared from ethylene, $CH_2=CH_2$, by (1) treatment of ethylene with H_2O/H_2SO_4, followed by treatment with $K_2Cr_2O_7/H_2SO_4$.
   (j) Treatment of propene, $CH_3CH=CH_2$ with (1) H_2O/H_2SO_4, followed by $K_2Cr_2O_7/H_2SO_4$ gives propanoic acid, CH_3CH_2COOH.

14.3 What Are the Structures, Names and Properties of Ethers?
   (a) Ethanol and dimethyl ether are constitutional isomers.
   (b) The solubility of low-molecular-weight ethers in water is comparable to that of low-molecular-weight alcohols in water.
   (c) Ethers undergo many of the same reactions that alcohols do.

14.4 What Are the Structures, Names and Properties of Thiols?
   (a) The functional group of a thiol is the –SH (sulfhydryl) group.
   (b) The parent name of a thiol is the name of the longest carbon chain that contains the -SH group.
   (c) The S-H bond is nonpolar covalent.
   (d) The acidity of ethanethiol is comparable to that of phenol.
   (e) Both phenols and thiols are classified as weak acids.
   (f) The most common biological reaction of thiols is their oxidation to disulfides.
   (g) The functional group of a disulfide is the –S-S- group.
   (h) Conversion of a thiol to a disulfide is a reduction reaction.

14.5 What Are the Most Commercially Important Alcohols?

(a) Today, the major carbon sources for the synthesis of methanol are coal and methane (natural gas), both nonrenewable resources.

(b) Today the major carbon sources for the synthesis of ethanol are petroleum and natural gas, both nonrenewable resources.

(c) Intermolecular acid-catalyzed dehydration of ethanol gives diethyl ether.

(d) Conversion of ethylene to ethylene glycol involves (1) an oxidation to ethylene oxide, followed by acid-catalyzed hydration (addition of water) of ethylene oxide.

(e) Ethylene glycol is soluble in water in all proportions.

(f) A major use of ethylene glycol is as automobile antifreeze.

ANSWERS TO SELF-TEST QUESTIONS

MULTIPLE CHOICE

1. a
2. b
3. a
4. b, c

5. a
6. b

7. c
8. b
9. c
10. partial oxidation produces a, while complete oxidation produces c
11. c
12. b
13. b

14.

a. No reaction. Thiols are not basic

b. CH_3S^- + H_2O. Thiols are weakly acidic and can react with a strong base.

c. $H_3CS-Hg-SCH_3$. Thiols have a strong affinity to bind heavy metals, such as mercury

d. $H_3CS-SCH_3$. Thiols are easily oxidized in the air (in the presence of O_2)

COMPLETION

1. elimination
2. oxidized, aldehyde, carboxylic acid
3. alcohol > ether
4. 1, 2-propanediol > propanol > ethylmethylether > propane
5.

a.
```
CH3CHCHCH3,              CH3CHCHCH3
   |  |                     |  |
  H3C  C-H               H3C  C-OH
        ||                     ||
        O                      O
  initial product         final product
```

b. CH3CH2OH

c.

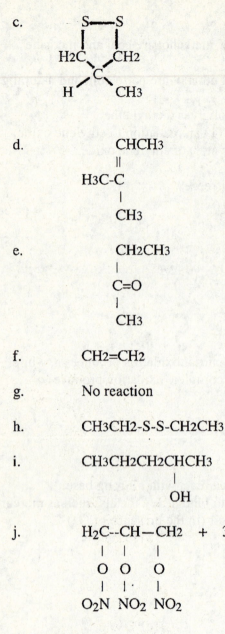

d.

          CHCH3
           ‖
    H3C-C
           |
          CH3

e.

          CH2CH3
           |
          C=O
           |
          CH3

f.   CH2=CH2

g.   No reaction

h.   CH3CH2-S-S-CH2CH3

i.   CH3CH2CH2CHCH3
                       |
                      OH

j.   H₂C--CH—CH₂   +   3 H2O
           | | |
           O   O   O
           | | |
          O₂N NO₂ NO₂

6. a.   neither
    b.   ether
    c.   alcohol
    d.   alcohol
    e.   alcohol
    f.   both an alcohol and an ether

7. a.   1, 2, 4-Butanetriol
    b.   Ethylisopropylether
    c.   3, 3-Dimethyl-2-butanol
    d.   Menthol or 2-Isopropyl-5-methylcyclohexanol
    e.   Propanethiol

8. a. $CH_3CH_2CH_2CH_2\text{-O-}CH_3$

    b. $CH_3CH_2CH_2C=CHCH_2CH_2\text{-OH}$
        $|$
        CH_2CH_3

    c. $H_2C\text{-CHCH}_2CH_2CH_3$
        $|\ \ |$
        HO OH

    d. Ethylene oxide

    e. Ethylene glycol (1, 2 Ethanediol)

        $CH_2\text{---}CH_2$
        $|\ \ \ \ |$
        HO   OH

9. oxidizing, reducing, oxidized
10. of the "bond strain" in the three-membered ring.

QUICK QUIZ

14.1
(a) True	(b) True	(c) False	(d) True
(e) True	(f) True	(g) True	(h) False

14.2
(a) True	(b) True	(c) False	(d) False
(e) False	(f) True	(g) True	(h) False
(i) True	(j) False		

14.3
(a) True	(b) False	(c) False

14.4
(a) True	(b) True	(c) True	(d) True
(e) True	(f) True	(g) True	(h) False

14.5
(a) True	(b) True	(c) True	(d) True
(e) True	(f) True		

167

15

Chirality: The Handedness of Molecules

CHAPTER OBJECTIVES

After you have studied the chapter and worked the assigned exercises in the text and the study guide, you should be able to do the following.

1. Distinguish the different types of isomers that have been presented in this and previous chapters and indicate an example of each.

2. Describe how connectivity (of the atoms) distinguishes stereoisomers from constitutional isomers.

3. Determine whether a molecule contains a stereocenter.

4. Determine the number of stereocenters in a chiral molecule and the maximum number of stereoisomers possible, by using the 2^n rule.

5. Distinguish between enantiomers and diastereomers.

6. Indicate the relationship between a racemic mixture and a pair of enantiomers.

7. Become proficient in using the R, S system to assign an R or S configuration at each stereocenter.

8. Explain how plane-polarized light and a polarimeter are used to experimentally characterize
an enantiomer.

9. Distinguish between the terms optically active, dextrorotatory, levorotatory and specific rotation.

10. Explain the importance of enantiomers in stereospecific interactions with enzymes in biological reactions and with the effectiveness of drug interactions.

11. Define the important terms and comparisons in this chapter and give specific examples where appropriate.

IMPORTANT TERMS AND COMPARISONS

Stereoisomerism

Stereocenter

Chiral and Achiral Molecules

Racemic Mixture

Diastereomers

R, S Priorities For Atoms and Groups

Chiral Drugs

Optically Active Molecule

Dextrorotatory (Clockwise; to the right)

Chirality of Biomolecules

Enantiomer and Enantiomeric Pair

Nonsuperposable Mirror Image and
   Superposable Mirror Image

Stereoisomers and Constitutional Isomers

R and S Configurations

2^n -Maximum Number of Stereoisomers

Plane Polarized Light and a Polarimeter

Specific Rotation, $[\alpha]$

Levorotatory (Counterclockwise; to the left)

Enzyme, Binding Site

FOCUSED REVIEW

Before we finish reviewing the major functional groups in organic chemistry and embark into the subject of **biochemistry**, the important concept of chirality is presented. It will become increasingly clear that this concept, and the other chemical concepts presented in previous chapters will provide the foundation for an understanding of biochemistry. As you encounter these basic concepts, do not be shy about returning to these chapters to review and reclaim a clear understanding of any vague terms. For example, a few of the concepts and reactions that were presented in previous chapters and are relevant to this chapter are presented below. You may want to extend this summary and prepare a similar review after each subsequent chapter.

Types of Isomers

Recall that **constitutional isomers** are molecules that have the same molecular formula, but different structural formulas (i. e., different order of attachment of their atoms). As a result of the structural differences, the chemical and physical properties of the isomers are different. A simple example of this is the molecular formula, C_2H_6O, and the different molecules (isomers) with this same formula. This formula may represent either (a) dimethyl ether or (b) ethanol.

(a) H_3COCH_3 (b) H_3CCH_2OH

These two isomers have different **structures**, different **types of bonds**, different **functional groups** and correspondingly different chemical and physical properties. These isomers are often also referred to as **structural isomers** or functional (group) isomers.

169

Although these constitutional isomers are in many cases very obvious, other isomer types are more subtle. **Stereoisomers** are compounds which have the same connections between atoms (i. e., each isomer has the same number and types of bonds), but have a different three-dimensional arrangements in space. There are two major classes of stereoisomers of interest to us.

The first type are the **stereoisomers** that results from the restriction of rotation about a bond. The most obvious type of isomers in this case may be the *cis-* and *trans*-isomers of a substituted olefin, such as $C_2Cl_2H_2$.

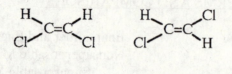

Cis-isomer   Trans-isomer

The second type is the **optical isomer,** which contains one or more stereocenters. Plane polarized light in a polarimeter is rotated on passing through a solution of an optical isomer. Plane polarized light does not rotate when it passes through a racemic mixture. Two subclasses of optical isomers are (a) enantiomers and (b) diastereomers.

In summary, the isomeric forms of molecules can be subdivided into groups as shown below. Much of our interest in this chapter will focus on simple organic molecules that are enantiomers or diastereomers. However, our interest in living cells will focus interest on biological molecules, including monosaccharides, amino acids, nucleotides, lipids and the major macromolecules such as polysaccharides, proteins and nucleic acids - all of which are chiral molecules and exhibit optical activity.

Isomer Classification

Isomer Type	Character	Examples
Constitutional	Same molecular formula, but different molecular structure	H_3CCH_2OH H_3COCH_3
Stereoisomers	Same connectivity of atoms, but different three-dimensional arrangement in space	
a. *Cis-Trans* isomers	*cis-* or *trans*-arrangement about a carbon-carbon double bond, such as observed in alkenes (Chapter 12)	*cis-* and *trans-* isomer of $C_2Cl_2H_2$
b. Optical		
(i) Enantiomers	Nonsuperimposable mirror images	D- and L-forms of glyceraldehyde
(ii) Diastereomers	Stereoisomers that are not enantiomers	α- and β-forms of D-glucose

MULTIPLE CHOICE. Select the correct answer from the choices listed. In some cases, more than one answer is correct.

1. Indicate which of the following molecules is chiral or achiral. If chiral, indicate how many stereocenters there are in the molecule.

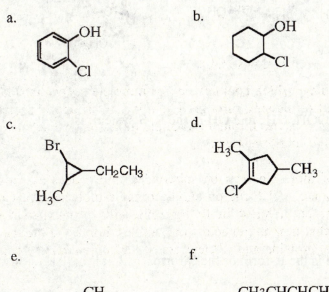

a.

b.

c.

d.

e.

f.

CH3CHCHCH2CH3
$\quad$ | $\quad$ |
HO OH

2. Enantiomers have which of the following characteristics?

a. rotate ordinary light
b. have the same melting point
c. have the identical molecular weight
d. are superimposable mirror images

3. A racemic mixture exhibits which of the following characteristics?

a. does not rotate plane polarized light
b. rotates plane polarized light
c. contains equal concentrations of both enantiomers
d. often contains diastereomers

4. Indicate which of the molecules in question 1 will have an enantiomer.

5. Indicate which of the following molecules are enantiomers and which ones are diastereomers.

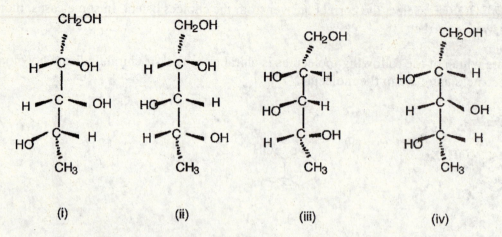

(i)   (ii)   (iii)   (iv)

6. The priority of the groups, CH3, COOH, NH2 and OH in the R, S system, is

    a.   CH3 > COOH > NH2 > OH
    b.   OH > NH2 > COOH > CH3
    c.   COOH > CH3 > NH2 > OH
    d.   COOH > OH > NH2 > CH3

7. Indicate whether the isomer shown is the R-form or the S-form.

a.

b.

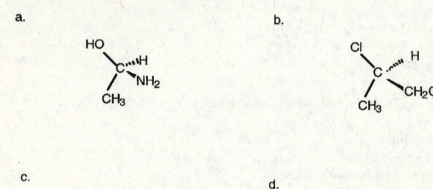

c.

d.

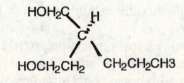

172

8. Calculate the maximum number of isomers in the molecules shown in questions 1 and 7.
   Molecules from <u>question 1</u>

   1a   1b

   1c   1d

   1e   1f

   Molecules from <u>question 7</u>

   7a   7b

   7c   7d

<u>COMPLETION.</u> Complete the following statements with the appropriate word, phrase or number.

1. Enantiomers are _____ mirror images. Because each enantiomer has at least one stereocenter, enantiomers must occur in _____.

2. The maximum number of stereoisomers that can occur for the molecule, $CH_3CH(OH)CH(OH)CH(OH)CHO$, is _____.

3. Diastereomers are _____, but they are not ____ ____ of each other. While the melting points for enantiomers are the same, the melting points for diastereomers are _____.

4. Thalidomide is perhaps the most notorious drug in recent history and one that involves enantiomers. To reduce cost, the drug was sold as a racemic mixture of the enantiomeric pair. Although this antidepressant drug was not approved for use in the United States, it was widely used in Canada and Europe. One enantiomer was very effective as an antidepressant. However, pregnant women who took the drug were also subjected to the other enantiomer, which was mutagenic (causes changes in DNA) and antiabortive. As a result, many women who took the drug gave birth to severely deformed children. The enantiomeric pair is shown below.

         Indicate how many stereocenters there are and identify the R- and the S-forms.

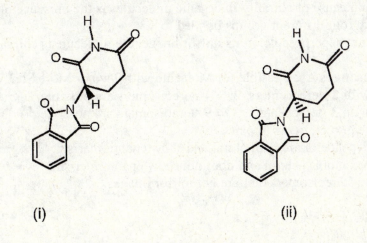

(i)   (ii)

5. Many biological molecules found in living systems are chiral. The steroid sex hormones are derived from cholesterol. Testosterone is responsible for the development of the secondary sex characteristics of males, while estrogens, such as estradiol, are required for the secondary sexual characteristics of females.

Indicate the number of stereocenters in testosterone and estradiol and the maximum number of stereoisomers possible for each.

Indicate which molecule contains an aromatic ring.

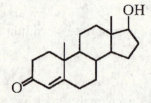

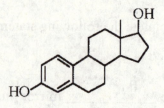

testosterone   estradiol

QUICK QUIZ

Indicate whether the statements are true or false. The number of the question refers to the section of the book that the questions were taken from.

15.1 What Is Enantiomerism?
   (a) The *cis* and *trans* stereoisomers of 2-butene are achiral.
   (b) The carbonyl carbon of an aldehyde, ketone, carboxylic acid, or ester cannot be a stereocenter.
   (c) Stereoisomers have the same connectivity of their atoms.
   (d) Constitutional isomers have the came connectivity of their atoms.
   (e) An unmarked cube is achiral
   (f) A human foot is chiral..
   (g) Every object in nature has a mirror image.
   (h) The most common cause of chirality in organic molecules is the presence of a tetrahedral carbon atom with four different groups bonded to it.
   (i) If a molecule is not superposable on its mirror image, the molecule is chiral.

15.3 How Many Stereoisomers Are Possible for Molecule with Two or More Stereocenters?
   (a) For a molecule with 2 stereocenters, $2^2 = 4$ stereoisomers are possible.
   (b) For a molecule with 3 stereocenters, $3^2 = 9$ stereoisomers are possible.
   (c) Enantiomers, like gloves, occur in pairs.
   (d) 2-Pentanol and 3-pentanol are both chiral and show enantiomerism.
   (e) 1-Methyl-cyclohexanol is achiral and does not show enantiomerism.
   (f) Diastereomers are stereoisomers that are not mirror images.

174

15.4 What Is Optical Activity and How Is Chirality Detected in the Laboratory?
   (a) If a chiral compound is dextrorotatory, its enantiomer is levorotatory by the same number of degrees.
   (b) A racemic mixture is optically inactive.
   (c) All stereoisomers are optically active.
   (d) Plane-polarized light consists of light waves vibrating in parallel planes

ANSWERS TO SELF-TEST QUESTIONS

MULTIPLE CHOICE

1.
a.	achiral, 0		b.	chiral, 2	
c.	chiral, 3		d.	chiral, 1	
e.	achiral		f.	chiral, 2	

2.   b, c
3.   a, c
4.   all chiral molecules - b, c, d & f
5.   enantiomers;   i & iii; ii & iv
     diastereomers;   i & ii; i & iv; ii & iii; iii & iv
6.   b
7.   a.   R
     b.   R
     c.   S
     d.   S
8.   From question 1

a.	0		b	4
c.	8		d.	2
e	0		f.	4

From question 7

a.	2		b.	2
c.	2		d.	2

COMPLETION

1.   nonsuperimposable, pairs
2. $2^3 = 8$
3.   stereoisomers, mirror images, different
4.   There is one stereocenter in each enantiomer.
     (i) is R-thalidomide; (ii) is S-thalidomide
5.   Testosterone has 6 stereocenters and a maximum of 64 possible stereoisomers
     Estradiol has 5 stereocenters and a maximum of 32 possible stereoisomers.
     Estradiol contains an aromatic ring.

QUICK QUIZ

15.1

(a) True	(b) True	(c) True	(d) False
(e) True	(f) True	(g) True	(h) True
(i) True			

15.3

(a) True	(b) False	(c) True	(d) False
(e) True	(f) True		

15.4

(a) True	(b) True	(c) False	(d) True

"Common sense is the collection of
prejudices acquired by age eighteen"
        - Albert Einstein

16

Amines

CHAPTER OBJECTIVES

After you have studied the chapter and worked the assigned exercises in the text and the
study guide, you should be able to do the following.

1. Write out a general structural formula for a primary, secondary and tertiary amine and for a
   quaternary ammonium salt.

2. Write out the names for simple amines, including aliphatic, cyclic and aromatic
   compounds.

3. Rationalize the trend in the boiling points and water solubility for simple primary amines.

4. Describe the hydrogen bonding that takes place (a) between two amines and (b) between
   amines and water and discuss the influence of hydrogen bonding on the physical properties
   of amines.

5. Give the names and structures for natural and synthetic amines that are used medicinally.

6. Write out the chemical equations for the reaction of amines with inorganic and organic
   carboxylic acids (RCOOH).

7. Describe how the presence of an aromatic or aliphatic group in an amine influences its
   base strength.

8. Define the important terms and comparisons in this chapter and give specific examples
   where appropriate.

IMPORTANT TERMS AND COMPARISONS

Aromatic and Aliphatic Amines   Primary, Secondary and Tertiary Amines

Heterocyclic Amines, Heterocyclic Aliphatic Amines and Heterocyclic Aromatic Amines

Amphetamines (Pep Pills)   Alkaloids

Quaternary Ammonium Ion   Hydrogen Bond Donors and

Organic Bases   Hydrogen Bond Acceptors

N-Substituted Amines K_b and pK_b

Acid/Base Chemistry of Amines   Benzodiazepine and Tranquilizers

SELF-TEST QUESTIONS

MULTIPLE CHOICE. Select the correct answer from the choices listed. In some cases, more than one choice will be correct.

1. The name of the compound given below is:

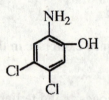

    a.   2-Amino-4, 5-dichlorophenol

    b.   Dichlorohydroxylaniline

    c.   2-Amino-4, 5-dichlorocyclohexanol

    d.   3, 4-Dichloro-1-hydroxylaniline

2. The structure of N,N-Dimethylaniline is:

a.   b.

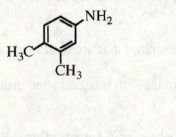 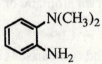

c.   d.

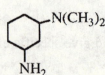

178

3. Protonation of amines that are used as drugs has the advantage that it

    a. increases solubility   c. decreases reduction
    b. inhibits oxidation   d. increases basic character

4. The pKb or Kb value for some basic amines are given below. Indicate which amine is the strongest base.

    a. pKb = 4.5   c. pKb = 7.0
    b. Kb = 1 x 10^{-8} d. Kb = 2 x 10^{-3}

5. Indicate which compound would be expected to have the highest boiling point.

    a. CH3CH2CH2OH   c. CH3CH2OCH3

    b. CH3CH2CH2NH2   d. CH3NHCH2CH3

6. Benzodiazepine, dopamine, epinephrine, procaine are all examples of:

    a. chiral molecules   c. aromatic compounds
    b. alkaloids   d. synthetic pain killers

7. A potential drug is made from a newly synthesized aromatic amine (pKb = 8.4). Calculate the ratio of [amine/protonated amine] that will be present in the blood (pH = 7.4). You may use the Henderson-Hasselbalch equation.

    a. 100/1   c. 50/1
    b. 90/1   d. 10/1

COMPLETION. Answer or complete the questions as indicated.

1. Indicate which compounds is an aliphatic amine, aromatic amine, heterocyclic amine, aromatic heterocyclic amine or a combination of these, or none of these.
    Indicate also whether it is a primary, secondary or tertiary amine.

179

a.

CH₂NH₂

b.

N-H

c.

N-H
N-CH₂CH₃

d.

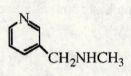
N
CH₂NHCH₃

e.

H
N

NH₂

f.

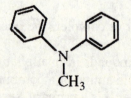

2. Draw the structures for these molecules.

    a. Ethylpropanamine
    b. Cyclohexanamine
    c. N-Methylaniline
    d. 1-Amine-4-chlorocyclohexane
    e. 5-Methylheptanamine
    f. Diphenylamine

3. Name these compounds

    a.

    NH₂
    CH₃

    b.

    NH₂

    c.

    H2NCH2CH2CH2CH2NH2

    d

    N
    CH₃

180

4. Indicate the order of increasing solubility of these compounds in water.

NH_3, NH_2CH_3, $N(CH_3)_2(C_6H_5)$

_____ < _____ < _____

5. Complete the equations for the following acid-base reactions. If no reaction occurs, indicate no reaction.

a. $(CH_3)_3N$ + HCl $\longrightarrow$

b. $(CH_3CH_2)_2H_2^+$ + OH^- $\longrightarrow$

c. $N(CH_3)_4^+ Cl^-$ + NaOH $\longrightarrow$

6. a. Secondary amines have _____ hydrogen(s) bound directly to the nitrogen.

   b. The compound $[(CH_3CH_2)_3NH]^+Br^-$ is a _____ and can be made by the reaction of _____ with _____.

MATCHING. Match the term in the right-hand column with the most appropriate answer in the left column.

1.	Weakest class of amine bases	a.	"Poison hemlock"
2.	Amphetamine	b.	Tranquilizer
3.	Pyridine	c.	Basic nitrogen-containing compounds extracted from plants
4.	Methylamine	d.	Pep pills, action similar to that of epinephrine
5.	Librium or Valium	e.	Water-soluble ammonium salt
6.	MethadoneHCl	f.	Aromatic amine
7.	Alkaloid	g.	Heterocyclic aromatic amine
8.	(S)-Coniine	h.	Aliphatic amine

Indicate whether the statements are true or false. The number of the question refers to the section of the book that the question was taken from.

16.1 What Are Amines?
(a) *tert*-Butylamine is a 3° amine.
(b) In an aromatic amine, one or more of the groups bonded to nitrogen is an aromatic ring.
(c) In a heterocyclic amine, the amine nitrogen is one of the atoms of a ring.
(d) The Lewis structures of both NH_4^+ and CH_4 show the same number (eight) of valence electrons, and the VSEPR model predicts tetrahedral geometry for each.
(e) There are four constitutional isomers with the molecular formula C_3H_9N.

16.2 How Do We Name Amines?
(a) In the IUPAC system, primary aliphatic amines are named as alkanamines.
(b) The IUPAC name of $CH_3CH_2CH_2CH_2CH_2NH_2$ is 1-pentylamine.
(c) 2-Butanamine is chiral and shows enantiomerism.
(d) *N,N*-Dimethylaniline is a 3° aromatic amine.

16.3 What Are the Physical Properties of Amines?
(a) Hydrogen bonding between 2° amines is stronger than that between 2° alcohols.
(b) Primary and secondary amines generally have higher boiling points than hydrocarbons with comparable carbon skeletons.
(c) The boiling points of amines increase as the molecular weight of the amine increases.

16.4 How Do We Describe the Basicity of Amines?
(a) Aqueous solutions of amines are basic.
(b) Aromatic amines such as aniline are in general weaker bases than aliphatic amines such as cyclohexanamine.
(c) Aliphatic amines are stronger bases than inorganic bases such as NaOH and KOH.
(d) Water-insoluble amines react with strong aqueous acids such as HCl to form water-soluble salts.
(e) If the pH of an aqueous solution of a 1° aliphatic amine, RNH_2, is adjusted to pH 2.0 by the addition of concentrated HCl, the amine will be present in solution almost entirely as its conjugate acid, RNH_3^+.
(f) If the pH of an aqueous solution of a 1° amine, RNH_2, is adjusted to pH 10.0 by the addition of NaOH, the amine will be present in solution almost entirely as the free base, RNH_2.
(g) For a 1° aliphatic amine, the concentrations of RNH_3^+ and RNH_2 will be equal when the pH of the solution is equal to the pKb of the amine.

ANSWERS TO SELF-TEST QUESTIONS

MULTIPLE CHOICE

1. a   4. d

2. b
3. a, b

5. a
6. c
7. d

COMPLETION

1.
   a.   A primary, aliphatic amine (N is attached to only an aliphatic carbon)

   b.   A secondary, aliphatic heterocyclic amine (contains no aromatic groups)

   c.   A heterocyclic ring, with two aliphatic amines, one is a secondary amine, while the other is a tertiary amine

   d.   A secondary, aliphatic amine and a heterocyclic, aromatic amine

   e.   A primary, aromatic amine and a secondary, heterocyclic aromatic amine

   f.   a tertiary, aromatic amine

2.

   a.

        CH3CH2NHCH2CH2CH3

   b.

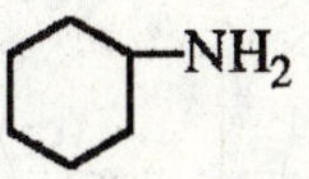

   c.

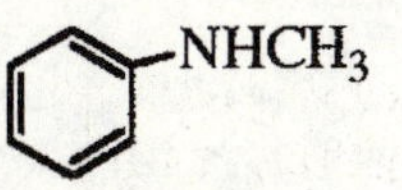

   d.

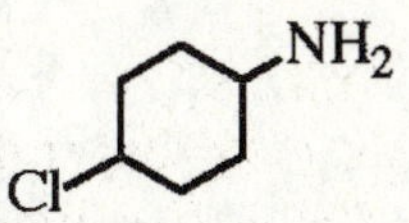

   e.

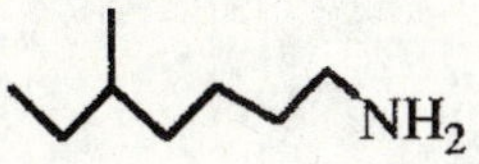

   f.

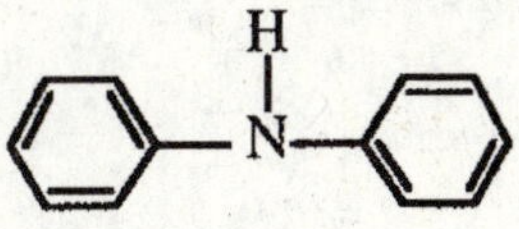

3.
   a.   o-Methylaniline (o-Toluidine)

   b.   Cyclohexanamine or cyclohexylamine

   c.   1, 4-Butandiamine

   d.   Methyldiphenylamine

4.
   a. $N(CH_3)_2(C_6H_5) < NH_2CH_3 < NH_3$

5.
   a. $(CH_3)_3N + HCl \longrightarrow (CH_3)_3NH^+ Cl^-$

   b. $(CH_3CH_2)_2H_2^+ + OH^- \longrightarrow (CH_3CH_2)_2NH + H_2O$

   c. $N(CH_3)_4^+ + NaOH \longrightarrow$ No reaction

6.
   a.   one

   b.   quaternary amine or quaternary ammonium salt.
        Reaction of $(CH_3CH_2)_3N + HBr$

183

MATCHING

1.	f	5.	b
2.	d	6.	e
3.	g	7.	c
4.	h	8.	a

QUICK QUIZ

16.1

(a) False   (b) True   (c) True   (d) True

(e) True

16.2

(a) True   (b) False   (c) True   (d) True

16.3

(a) False   (b) True   (c) True

16.4

(a) True   (b) True   (c) False   (d) True

(e) True   (f) True   (g) True

Man's mind stretched by a new idea
never goes back to its original dimensions.
    - paraphrase from Oliver Wendell Holmes

17

Aldehydes and Ketones

CHAPTER OBJECTIVES

After you have studied the chapter and worked the assigned exercises in the text and the study guide, you should be able to do the following.

1. Write out the characteristic functional group for an aldehyde and a ketone and contrast these to the structure and bond types found in functional groups presented in previous chapters.

2. Given the structural formula for an aldehyde or ketone, name the compound with the IUPAC nomenclature.

3. Given the IUPAC or common name for an aldehyde or ketone, write its structural formula.

4. Choosing organic compounds of approximately the same molecular weight, compare the physical properties - boiling point and solubility in water - for an alkane, alcohol, amine, ketone, aldehyde, and ether.

5. Referring back to Chapter 14, write out reactions for the preparation of aldehydes and ketones that are derived from the oxidation of alcohols that contain 2, 3 and 4 carbons.

6. Write out the balanced reactions for a positive Tollens' test (silver-mirror test).

7. Describe the basis for using the Tollin's test to distinguish between an aldehyde and a ketone.

8. Clearly write out the general **structures** for a hemiacetal and an acetal that are derived from the addition reaction of an alcohol with an aldehyde. Provide at least one specific example of each.

9. Draw the general **structures** for a hemiacetal and an acetal that are derived from the addition reaction of an alcohol with a ketone. Provide at least one specific example of each.

10. Write the **chemical reactions** for the formation of a hemiacetal and an acetal that are derived from an addition reaction with an aldehyde. Provide examples using at least two different aldehydes.

11. Write the **chemical reactions** for the formation of a hemiacetal and an acetal that are derived from an addition reaction with a ketone. Provide examples using at least two different ketones.

12. Using a single molecule that contains both an alcohol and an aldehyde (ketone) functional group, show how a cyclic hemiacetal, and then on further reaction, a cyclic acetal, is formed.

13. Write out the equations for the reduction of aldehydes and ketones to produce a primary and secondary alcohol, respectively.

14. Summarize the oxidation, reduction and addition reactions for aldehydes and ketones.

15. Indicate what a ketone molecule must contain to undergo a keto-enol tautomerization.

16. Indicate at least two different examples of ketones that can occur as tautomers and draw their keto and enol forms

17. Define the important terms and comparisons in this chapter and give specific examples where appropriate.

IMPORTANT TERMS AND COMPARISONS

Carbonyl Group

Saturated and Unsaturated Aldehydes

Sufixes -al and –one

Tollens' Test - the Silver Mirror Test

Hemiacetal and Acetal (From a Ketone)

Keto and Enol Forms

Reducing Agents - NaBH4 and NADH

Reactions with Aldehydes
    Oxidation to a Carboxylic Acid
    Reduction to a Primary Alcohol
    Addition of Alcohol to Form:
        Hemiacetal or Acetal

Aldehyde and Ketone

IUPAC Nomenclature

Primary and Secondary Alcohols

Hemiacetal and Acetal (From an Aldehyde)

Linear and Cyclic Acetals

Tautomerism and Tautomers

Hydride Anion (H^-)

Reactions with Ketones
    No Oxidation
    Reduction to a Secondary Alcohol
    Addition of Alcohol to Form:
        Hemiacetal or Acetal

In Chapters 10-17, we have increasingly expanded the coverage of organic chemistry - the study of carbon-containing compounds with various functional groups (i. e., reactive groups), with an emphasis on the characteristic physical properties and chemical reactions associated with each group. The major functional groups presented to this point are the following.

MAJOR FUNCTIONAL GROUPS

I. Contains Only C and H
(Hydrocarbons)

A. Saturated

Alkanes $-\overset{|}{\underset{|}{C}}-\overset{|}{\underset{|}{C}}-$

B. Unsaturated

Alkene $\underset{}{>}C=C<$

Alkyne $-C \equiv C-$

C. Aromatic

II. Contains C, H and O

A. Carbonyl Groups

Aldehyde $R-\overset{O}{\overset{||}{C}}H$

Ketone $R-\overset{O}{\overset{||}{C}}-R'$

B. Others

Alcohol   R-OH

Ether   R-O-R'

Phenol   ⬡—OH

III. Contains C, H and S or N

Thiol or Mercaptan   R-SH   Disulfide   R-S-S-R'
Amines   NHnR3-n

Organize and review the summary sheets of the reactions that characterize each functional group. The importance of making and continually reviewing (even casually) these sheets throughout the course will become vividly clear when you carry out a major review of the material for each hourly exam and the final exam.

This chapter focuses on aldehydes and ketones, their physical properties and the chemical reactions associated with each functional group. The production of hemiacetals and acetals from the reaction of an aldehyde or a ketone with an alcohol is of special importance here and in the subsequent chapter on carbohydrate chemistry. The following table summarizes these general structures. You should be able to readily recognize and distinguish between these six compounds before going onto the SELF-TEST.

Aldehyde		**Hemiacetal**	**Acetal**
O		OH	OR'
‖		|	|
R-C-H	⟶	**R-C-H**	**R-C-H**
		|	|
		OR'	OR'

187

Ketone		Hemiacetal		Acetal
O ‖ **R-C-R''**	⟶	OH | **R-C-R''** | OR'		OR' | **R-C-R''** | OR'

Consider the reaction of an adehyde or a ketone with an alcohol to produce the corresponding hemiacetal or acetal shown in the table above. Note that the **bold-type groups** remain in the structure for the three related compounds (reactant and either product). The R primed group (R') comes from the alcohol group that reacts with the aldehyde or ketone. Write out the reactions for different alcohols, reacting with an aldehyde or ketone, showing the structure of the product. Indicate from which reactant each of the R groups is derived.

SELF-TEST QUESTIONS

<u>MULTIPLE CHOICE</u>. Select the correct answer or answers from the choices listed. In some cases, more than one answer is correct.

1. Mild oxidation of an aldehyde produces a carboxylic acid. However, mild oxidation of a ketone results in:

    a.   an ether   c.   carboxylic acid
    b.   secondary alcohol   d.   no reaction

2. Identify which of the following compounds are an aldehyde, hemiacetal or an acetal.

   a.   b.

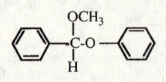

   c.   d.

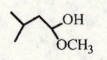
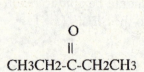

   e.   f.

$$O$$
$$\|$$
$$CH_3CH_2\text{-}C\text{-}CH_2CH_3$$

3. Select compounds in question 2 that have a carbonyl group.

4. The reaction of an alcohol with a ketone is an example of a(n):

   a. elimination reaction   c. hydrolysis reaction
   b. addition reaction   d. substitution reaction

5. Ketones are prepared by the oxidation of:

   a. secondary alcohol   c. tertiary alcohol
   b. primary alcohol   d. all of these

6. The characteristics that are produced in a positive Tollens' test are:

   a. Oxidation of an aldehyde and reduction of $Ag(NH_3)_2^+$.
   b. Reduction of an aldehyde and oxidation of $Ag(NH_3)_2^+$.
   c. Oxidation of a ketone and production of a silver mirror.
   d. All of the above.

7. The structural formula for benzaldehyde is:

   a.   b.

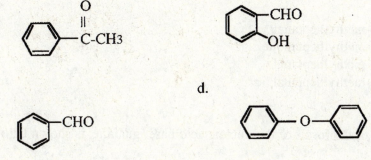

   c.   d.

8. Indicate which of the following molecules can undergo a keto-enol equilibrium.

   a. CH3CH2-OH

   b. CH3CH2CH2COCH3

   c.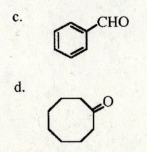

   d.

9. The structural formula for 2,3-Dimethylcyclohexanone is:

a.

b.

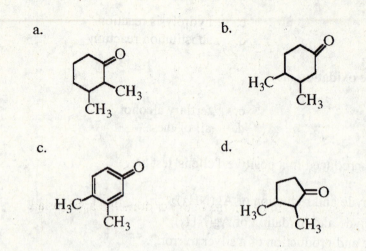

c.

d.

10. The IUPAC name for

$H_3CCCHCH_2CH_2CH_2CHO$ is:

with H₃C Br on top attached to the chain and Br below

a. 5, 6-Dibromo-6-methylheptanal
b. 1, 2-Dibromo-1-methylheptanone
c. 5, 6-Dibromo-1-methylheptanal
d. 1 ,2-Dibromo-1-methylheptanal

11. The structure for one of the forms of the nucleic acid base, guanine, is shown below. This form is:

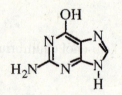

a. a ketone
b. the keto form of guanine
c. the enol form of guanine
d. an acetal

190

12. Indicate in which of the following molecules will the <u>product</u> of an oxidation or reduction (indicate) be an aldehyde or a ketone.

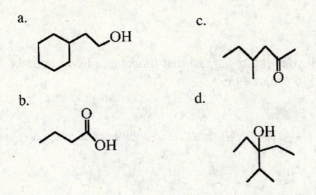

a.

c.

b.

d.

COMPLETION. Write the word, phrase or number in the blank or draw the appropriate structure in answering the question.

1. Write out the IUPAC names for the following molecules.

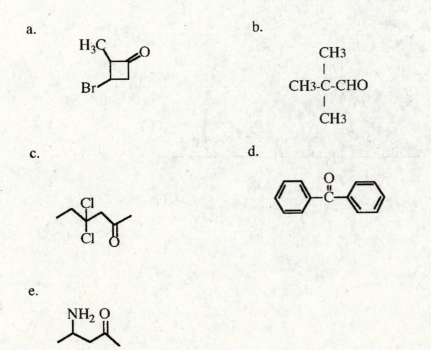

a.

b.

CH3
|
CH3-C-CHO
|
CH3

c.

d.

e.

2. Given compounds of comparable molecular weight, the order of decreasing boiling points for an alcohol, alkane, ether and aldehyde is:

_____ > _____ > _____ > _____

191

3. The common name for one of the simplest aldehydes, CH3CH, is _____.
$$\overset{\parallel}{\underset{O}{}}$$

4. The general structure of an aldehyde would suggest that this class of compounds
_____ hydrogen bond to each other.

5. The complete hydrolysis of an acetal yields an _____ and _____.

$$\overset{O}{\overset{\parallel}{}}$$

6. The reaction of a ketone, R-C-R", with two molecules of an alcohol, R'-OH, yields an
_____ , which contains two ether linkages. Draw the general structure below.

_____

7. Complete the following equations (i. e., write the structure of the product). [H] and [O] indicate reduction and oxidation, respectively. If no reaction occurs, indicate no reaction.

a.

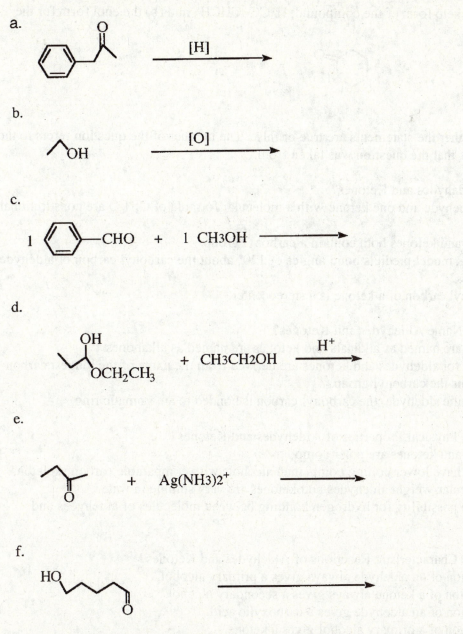

$\xrightarrow{[H]}$

b.

$\xrightarrow{[O]}$

c.

1 —CHO + 1 CH3OH $\longrightarrow$

d.

+ CH3CH2OH $\xrightarrow{H^+}$

e.

+ Ag(NH3)2$^+$ $\longrightarrow$

f.

$\longrightarrow$

8. The oxidation of (C₆H₅)-CH₂CHO yields a molecule with the structural formula

_____

9. In an reduction reaction of an aldehyde with NaBH4, a _____ anion, which is a hydrogen with _____ electrons, attacks the carbonyl _____ of the aldehyde functional group.

10. The mixing of an aldehyde with an acid or base yields _____ reaction because all aldehydes are _____ acidic or basic in character.

193

11. Draw the (a) keto form of the compound, $H_3CC=CHCH_3$ and (b) the enol form for the

$$\overset{\displaystyle OH}{\underset{\displaystyle |}{}}$$

compound

QUICK QUIZ
     Indicate whether the statements are true or false. The number of the question refers to the section of the book that the question was taken from.

17.1 What Are Aldehydes and Ketones?
     (a) The one aldehyde and one ketone with a molecular formula of C_3H_6O are constitutional isomers.
     (b) Aldehydes and ketones both contain a carbonyl group.
     (c) The VSEPR model predicts bond angles of $120°$ about the carbonyl carbon of aldehydes and ketones.
     (d) The carbonyl carbon of a ketone is a stereocenter.

17.2 How Do We Name Aldehydes and Ketones?
     (a) Aldehydes are named as alkanals and ketones are named as alkanones.
     (b) The names for aldehydes and ketones are derived from the name of the longest carbon that contains the carbonyl group.
     (c) In an aromatic aldehyde, the carbonyl carbon is bonded to an aromatic ring.

17.3 What Are the Physical Properties of Aldehydes and Ketones?
     (a) Aldehydes and ketones are polar compounds.
     (b) Aldehydes have lower boiling points than alcohols with comparable carbon skeletons.
     (c) Low-molecular-weight aldehydes and ketones are very soluble in water.
     (d) There is no possibility for hydrogen bonding between molecules of aldehydes and ketones.

17.4 What Are the Characteristic Reactions of Aldehydes and Ketones?
     (a) The reduction of an aldehyde always gives a primary alcohol.
     (b) The reduction of a ketone always gives a secondary alcohol.
     (c) The oxidation of an aldehyde gives a carboxylic acid.
     (d) The oxidation of a primary alcohol gives a ketone.
     (e) Tollens' reagent can be used to distinguish between an aldehyde and a ketone.
     (f) Sodium borohydride, $NaBH_4$, reduces an aldehyde to a primary alcohol.
     (g) The addition of one molecule of alcohol to the carbonyl group of a ketone gives a hemiacetal.
     (h) The reaction of an aldehyde with two molecules of alcohol gives an acetal plus a molecule of water.
     (i) The formation of hemiacetals and acetals is reversible.
     (j) The cyclic hemiacetal formed from 4-hydroxypentanal has two stereocenters and can exist as a mixture of $2^2 = 4$ stereoisomers.

194

17.5 What Is Keto-Enol Tautomerism?
   (a) Keto and enol tautomers are constitutional isomers.
   (b) For a pair of keto-enol tautomers, the keto form generally predominates.

ANSWERS TO SELF-TEST QUESTIONS

<u>MULTIPLE CHOICE</u>

1. d
2. a (acetal), c (aldehyde), e (hemiacetal)
3. c, f
4. b
5. a
6. a

7. c
8. b, d
9. a
10. a
11. c
12. a (oxidation to an aldehyde)
    b (reduction to an aldehyde)

<u>COMPLETION</u>

1. a. 3-Bromo-2-methylcyclobutan-1-one
    b. 2, 2-Dimethylpropanal
    c. 4, 4, Dichlorohexan-2-one
    d. Diphenylketone
    e. 4-Aminepentan-2-one
    (Answers a, c and e are alternate, acceptable forms to naming ketones)
2. alcohol > aldehyde > ether > alkane
3. acetaldehyde
4. does not
5. alcohol and aldehyde (or ketone)

$$OR'$$
$$|$$

6. acetal, R-C-R", in which the R' groups are derived from the alcohol
$$|$$
$$OR'$$

7.

a.

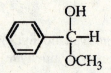

b.

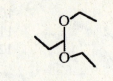 or

c.

d.

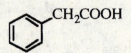

e.   no reaction

f.

8.

CH₂COOH

9.   hydride, 2, carbon

10.   no, neither

11.   O
           ‖
a. CH3CCH2CH3

b.

196

QUICK QUIZ

17.1

(a) True	(b) True	(c) True	(d) False

17.2

(a) True	(b) True	(c) True

17.3

(a) True	(b) True	(c) True	(d) True

17.4

(a) True	(b) True	(c) True	(d) False
(e) True	(f) True	(g) True	(h) True
(i) True	(j) True		

17.5

(a) True	(b) True

Somewhere, something incredible is
waiting to be known.
- Carl Sagan

18

Carboxylic Acids

CHAPTER OBJECTIVES

After you have studied the chapter and worked the assigned exercises in the text and the study guide, you should be able to do the following.

1. Write out the three ways to represent the functional group for a carboxylic acid and the representation for a carboxylate salt.

2. Indicate the requirements for a carboxylic acid to be classified as a fatty acid.

3. Draw the structure of the carboxylic acids that are 1) saturated and contain one to six carbons, 2) *cis*-unsaturated fatty acid, 3) a *trans*-fatty acid and 4) an omega-3 fatty acid.

4. Explain the influence of hydrogen bonding in the physical properties of carboxylic acids.

5. Explain how the degree of unsaturation affects the melting point of the fatty acid.

6. Write the reaction for the dissociation of a carboxylic acid and also for the reaction with a strong base, such as NaOH.

7. Given the common or IUPAC name for a carboxylic acid, write out its structural formula.

8. Given the condensed structural formula for a carboxylic acid, name the compound by either its common or IUPAC name.

9. Write out the complete equation for the:
    a.   reaction of a carboxylic acid with a base to form a carboxylate salt.
    b.   reaction of a carboxylic acid with an alcohol to form an ester.
    c.   reaction of a carboxylic acid with LiAlH4 to produce an alcohol.
    d.   decarboxylation reaction of a β-ketocarboxylic acid.

10. Indicate how soap and detergents are prepared and how they will differ when used in hard water.

11. Explain how micelles are formed and play a critical role in cleaning clothes.

12. Indicate an experimental strategy that could be effectively used to separate a mixture that contains a water-insouble alcohol and a carboxylic acid.

13. Define the important terms and comparisons in this chapter and give specific examples where appropriate.

IMPORTANT TERMS AND COMPARISONS

Carbonyl Group and Carboxyl Group
Hydrogen Bonding
Ka and pKa
Alpha (α), Beta (β) and Gamma (γ) Positions
Saturated and Unsaturated Fatty Acids
Hydrogenation of Unsaturated Fatty Acids
Saponification (Base-Promoted Hydrolysis)
Hard Water
Micelles
Linear Alkylbenzenesulfonates (LAS)
Bleaches
Esters
Reactions for Carboxylic Acids
    Acid-Base Reactions
    Reduction by LiAlH4
    Fisher Esterification
    Decarboxylation (Loss of CO_2)
        β-Ketoacids

Carboxylic Acids
Carboxylic Acid Dimer
Substituent Effects on Acidity
Suffixes: -oic,-dioic and -oate
Palmitic(16:0), Stearic (18:0)
    and Oleic (18:1) Acids
Soaps and Detergents
Emulsifying Agents
London Dispersion Forces
Foam Stabilizers
Optical Brighteners
Ester Flavoring Agents
Health Related Terms
    Trans Fatty Acids
    Saturated Fatty Acids
    Monounsaturated Fatty Acids
    Polyunsaturated Fatty Acids
    LDL, HDL and LDL/HDL Ratio
    Omega-3 Fatty Acids
    Tricarboxylic Acid (TCA) Cycle
    Diabetes and Ketone Bodies

SELF-TEST QUESTIONS

MULTIPLE CHOICE. In the following exercises, select the correct answer from the choices listed.

1. The characteristic of fatty acids found in the cells of plants and higher animals are:

    a. The unsaturated fatty acids are nearly all *trans* fatty acids.
    b. Nearly all the saturated fatty acids have even number of carbons, including the carbon in the carboxylic acid functional group.
    c. Butanoic and hexanoic acids are among the most common saturated fatty acids.
    d. The most common saturated fatty acids contain between 14, 16, 18 or 20 carbons.

199

2. Carboxylate ion can exist in water at only:

   a. pH 7
   b. low pH

   c. high pH
   d. $[H^+]$ greater than 10^{-2} M

3. The saponification (hydrolysis of an ester in **basic** solution) of the carboxylic ester shown below yields the following products:

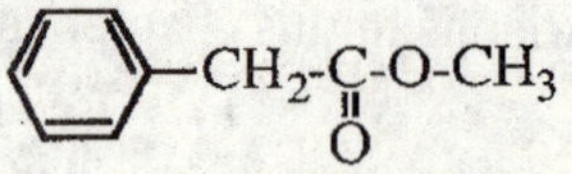

   a.

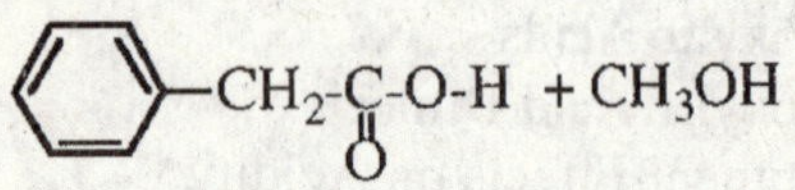

   b.

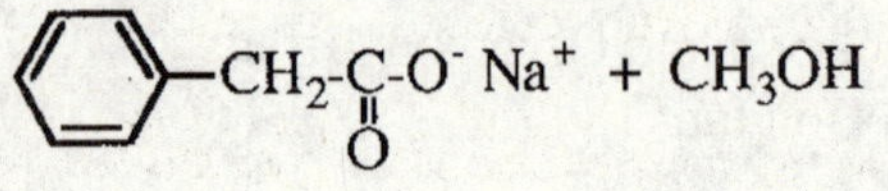

   c.

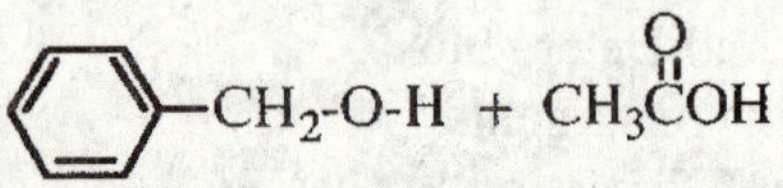

   d.

   None of these

4. Three organic acids are listed below with their pKa values.

		pK$_a$ value
Benzoic acid	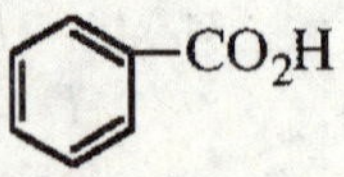 $-CO_2H$	4.19
Chloroacetic acid	$ClCH_2COOH$	2.86
Formic acid	$HCOOH$	3.74

   The order of decreasing acid strength is:

   a. chloroacetic acid > formic acid > benzoic acid
   b. benzoic acid > formic acid > chloroacetic acid
   c. all are the same
   d. not determinable from available data

5. The name of the compound, $H_3CCH_2CHCH_2CHC\text{-}OH$, is (with CH_3 and Br substituents, and a $C=O$)

   a. 1-Bromo-3-methylhexanoic acid
   b. 2-Bromo-4-methylhexanoic acid
   c. 2-Bromo-4-methylpentanoic acid
   d. 5-Bromo-3-methylpentanoic acid

6. Indicate which of the following acids is a β-keto acid.

   a. CH₃CCH₂CH₂COOH (with O double-bonded above first C)

   c. CH₃CCH₂COOH (with O double-bonded above first C)

   b. HO₂CCH₂CH₂CCH₂CH₃ (with O double-bonded above the second C)

   d. CH₃CH₂CCH₂OCH₃ (with O double-bonded above the C)

7. The product of the reduction of a carboxylic acid with LiAlH4 is a:

   a. primary alcohol
   b. ketone
   c. secondary alcohol
   d. aldehyde

8. Indicate which of the following are characteristic of unsaturated fatty acids.

   a. Found in greater abundance in animal fats
   b. If partially hydrogenated, lead to a mixture of *cis-* and *trans-*fatty acids
   c. They have a lower melting point than saturated fatty acids.
   d. They are usually liquid, while saturated fatty acids are usually solids.

9. The main force(s) that determine whether a fatty acid is liquid or solid is (are):

   a. London dispersion forces
   b. hydrogen bonding forces
   c. electrostatic interactions between charged groups
   d. all these forces

COMPLETION. Write the word, phrase or number in the blank or draw the appropriate structure in answering the questions.

1. In this table, indicate whether the following acids are carboxylic acids, fatty acids, saturated or unsaturated and how many carbons are in the acid. Indicate yes or no (or the # of Cs) in the appropriate column.

Acid	Carboxylic Acid	Saturated	Unsaturated	# of Cs
a. Hexanoic acid				
b. Linolenic acid (18:3)				
c. Acetic acid				
d. Myristic acid (14:0)				
e. H₃PO₄				

2. Name the functional group or groups contained in each of the following condensed structural formulas. Some of the functional groups were introduced in previous chapters.

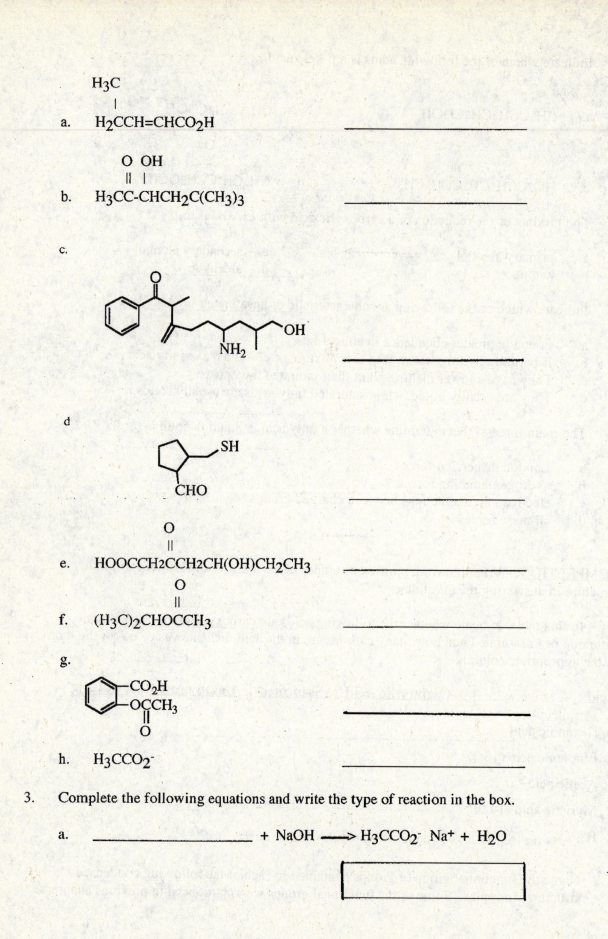

a.
$$\underset{\underset{\text{H}_2\text{CCH=CHCO}_2\text{H}}{\big|}}{\text{H}_3\text{C}}$$

_____

b.
$$\overset{\overset{\text{O}\ \ \text{OH}}{\underset{\text{|}}{\|}\ \ \underset{\text{|}}{\ }}{}\text{H}_3\text{CC-CHCH}_2\text{C(CH}_3)_3$$

_____

c.

_____

d

_____

e.
$$\text{HOOCCH}_2\overset{\overset{\text{O}}{\|}}{\text{C}}\text{CH}_2\text{CH(OH)CH}_2\text{CH}_3$$

_____

f.
$$(\text{H}_3\text{C})_2\text{CH}\overset{\overset{\text{O}}{\|}}{\text{C}}\text{CCH}_3$$

_____

g.

_____

h. H_3CCO_2^-

_____

3. Complete the following equations and write the type of reaction in the box.

a. _____ + NaOH ⟶ H_3CCO_2^- Na^+ + H_2O

┌─────────────────────────────┐
│   │
│   │
└─────────────────────────────┘

b. $CH_3CO_2^- \, Na^+ + HCl \longrightarrow$

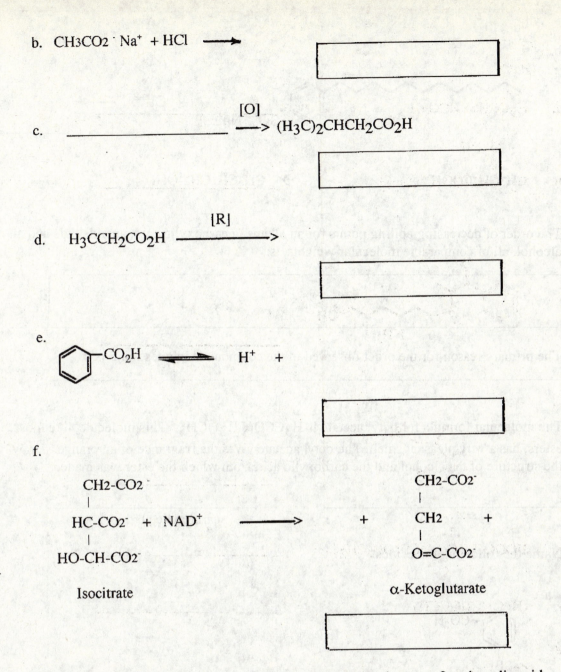

c. _____ $\xrightarrow{[O]}$ $(H_3C)_2CHCH_2CO_2H$

d. $H_3CCH_2CO_2H \xrightarrow{[R]}$

e.

benzene ring—CO_2H ⇌ $H^+ +$

f.

$$
\begin{array}{l}
CH2-CO_2^- \\
| \\
HC-CO_2^- \\
| \\
HO-CH-CO_2^-
\end{array}
+ NAD^+ \longrightarrow +
\begin{array}{l}
CH2-CO_2^- \\
| \\
CH2 \\
| \\
O=C-CO_2^-
\end{array}
+
$$

Isocitrate   α-Ketoglutarate

4. The structure of the dimer formed by hydrogen bonding between 2 carboxylic acid molecules, RCO_2H, is:

5. Insert the reactants or products in the following reactions.

a. $2 \, CH_3(CH_2)14COO^- \, Na^+ + Mg^{+2} \rightarrow$

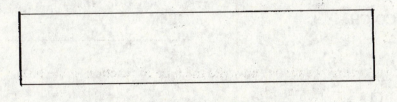

b.

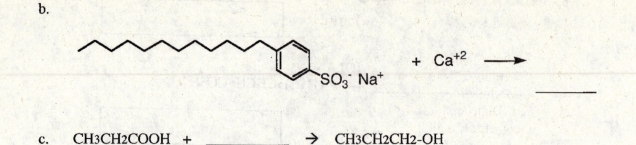

$+ \; Ca^{+2} \; \longrightarrow$

_____

c.   CH3CH2COOH + _____ → CH3CH2CH2-OH

6. The order of decreasing boiling points for an alkane, a carboxylic acid, an aldehyde and an alcohol, all of comparable molecular weights, is:

>	>	>	

7. The primary reason for the order observed in the compounds in question 5 is

_____.

8. The molecular formula for octyl acetate is $H_3CCO(CH_2)_7CH_3$ (with O double-bonded to the second C, shown as:

$$\overset{\displaystyle O}{\overset{\displaystyle \|}{}}$$

$H_3CCO(CH_2)_7CH_3$). This molecule, like most esters, has a very pleasant smell. The octyl acetate gives the fragrance of an orange. Draw the structure of the alcohol and the carboxylic acid from which the ester was made.

_____   _____

       alcohol   carboxylic acid

9. Name the following molecules.

a.

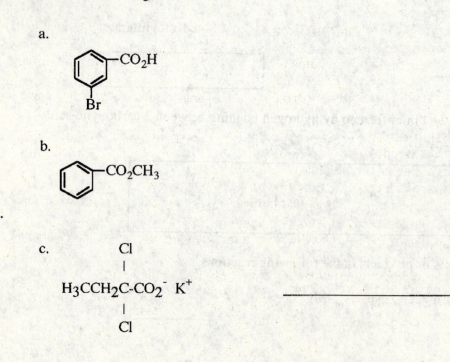

b.

—CO₂CH₃ (benzene ring with $-CO_2CH_3$ substituent)

c.   Cl
               |
   $H_3CCH_2C\text{-}CO_2^{-} \; K^{+}$ _____
               |
               Cl

10. Draw the structures for the following molecules.

    a.   2-Phenybutanoic acid
    b.   4-Amino-3-hydroxyoctanoic acid
    c.   o-Methylbenzoic acid
    d.   5, 5-Diaminohexanoic acid
    e.   Octanedioic acid

QUICK QUIZ

Indicate whether the statements are true or false. The number of the question refers to the section of the book that the question was taken from.

18.1 What Are Carboxylic Acids?

(a) The functional groups of a carboxylic acid are a carbonyl group and a hydroxyl group.
(b) The VSEPR model predicts bond angles of 180° about the carbonyl carbon of a carboxyl group.
(c) The VSEPR model predicts bond angles of 109.5° about the oxygen of the OH group of a carboxyl group.
(d) The carbonyl carbon of a carboxyl group can be a stereocenter, depending on its location within a molecule.
(e) Carboxylic acids can be prepared by chromic acid oxidation of primary alcohols and of aldehydes.
(f) The product of chromic acid oxidation of hexanoic acid is 1-hexanol.

18.2 How Do We Name Carboxylic Acids?

(a) The general name of an aliphatic carboxylic acid is alkanoic acid.
(b) A molecule containing two COOH groups is called a dicarboxylic acid.
(c) Ethanedioic acid (oxalic acid) is the simplest dicarboxylic acid.
(d) 3-Methylbutanoic acid is chiral.
(e) The simplest carboxylic acid is methanoic acid (common name: formic acid), HCO_2H.
(f) Benzoic acid is an aromatic carboxylic acid.
(g) Formic acid, which is the common name for HCO_2H, is derived from *formica*, the Latin name for ants.
(h) (S)-Lactic acid, $CH_3-CHOH-COOH$, contains two functional groups: a 2° alcohol and a carboxyl group.

18.3 What Are the Physical Properties of Carboxylic Acids?

(a) Carboxylic acids are polar compounds.
(b) The most polar bond of a carboxyl group is the C-O single bond.
(c) Carboxylic acids have significantly higher boiling points than aldehydes, ketones, and alcohols of comparable molecular weight.
(d) The lowest-molecular-weight carboxylic acids (formic, acetic, propanoic, and butanoic acids) are infinitely soluble in water.
(e) The order of decreasing boiling points for the following compounds is

205

(f) In order of increasing boiling point, the following compounds are

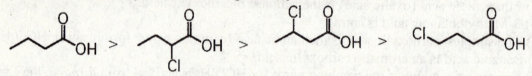

I   II   III   IV

18.4 What Are Soaps and Detergents?

(a) Fatty acids are long-chain carboxylic acids, with most consisting of between 12 to 20 carbons in an unbranched chain.

(b) An unsaturated fatty acid contains one or more carbon-carbon double bonds in its hydrocarbon chain.

(c) In most unsaturated fatty acids found in animal fats, vegetable oils, and biological membranes, the *cis* isomer predominates.

(d) In general, unsaturated fatty acids have lower melting points than saturated fatty acids with the same number of carbon atoms.

(e) Natural soaps are sodium or potassium salts of fatty acids.

(f) Soaps remove grease, oil, and fat stains by incorporating these substances into the nonpolar interior of soap micelles.

(g) "Hard water", by definition, is water that contains Ca^{2+}, Mg^{2+}, or Fe^{3+} ions, all of which react with soap molecules to form water-insoluble salts.

(h) The structure of synthetic detergents is patterned after that of natural soaps.

(i) The most widely used synthetic detergents are the linear alkylbenzenesulfonates (LAS).

(j) The present-day synthetic detergents do not form water insoluble salts with hard water.

18.5 What Are the Characteristic Reactions of Carboxylic Acids?

(a) Carboxylic acids are weak acids compared to mineral acids such as HCl, H_2SO_4, and HNO_3.

(b) Phenols, alcohols, and carboxylic acids have in common the presence of an –OH group.

(c) Carboxylic acids are stronger acids than alcohols but weaker acids than phenols.

(d) The order of acidity of the following carboxylic acids is:

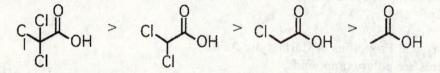

(e) The order of acidity of the following carboxylic acids is,

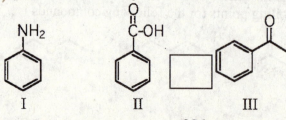

(f) The reaction of benzoic acid with aqueous sodium hydroxide gives sodium benzoate.

(g) If a mixture of the following three compounds is extracted from an organic solvent with a 1 M HCl aqueous solution, only the aromatic amine is extracted into the aqueous phase.

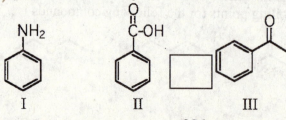

206

(h) Conversion of compound I to compound II can be accomplished by $K_2Cr_2O_7/H_2SO_4$.

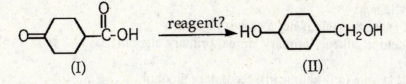

(I)   (II)

(i) The following ester can be prepared by treating benzoic acid with 1-butanol in the presence of a catalytic amount of H_2SO_4:

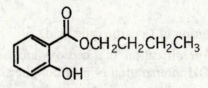

(j) Thermal decarboxylation of this β-ketoacid gives benzoic acid and carbon dioxide:

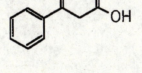

$CH_3CH_2CH_2\overset{O}{C}CH_2\overset{O}{C}OH$

(k) Thermal decarboxylation of this β-ketoacid gives 2-butanone and carbon dioxide.

ANSWERS TO SELF-TEST QUESTIONS

MULTIPLE CHOICE

1. b, d
2. c
3. b
4. a

5. b
6. c
7. a
8. b, c, d
9. a

COMPLETION

1.

Acid	Carboxylic Acid	Saturated	Unsaturated	# of Cs
a. Hexanoic acid	yes	yes	no	6
b. Linolenic acid (18:3)	yes	no	yes	18
c. Acetic acid	yes	yes	no	2
d. Myristic acid (14:0)	yes	yes	no	14

207

e. H_3PO_4 no   -   -   0

2. a.   alkene, carboxylic acid
    b.   ketone, alcohol (note that this is not a carboxylic acid)
    c.   aromatic ring, ketone, alkene, primary amine, primary alcohol
    d.   thiol, aldehyde
    e.   carboxylic acid, ketone (a β-keto acid), secondary alcohol
    f.   ester
    g.   carboxylic acid, carboxylic ester, aromatic ring
    h.   carboxylate anion

3. a. H_3CCO_2H; neutralization of (a carboxylic acid)
    b. H_3CCO2H + NaCl; acidification of a carboxylate salt by the addition of acid
    c. $(H_3C)_2CHCH_2CH_2OH$; preparation of a carboxylic acid by oxidation of a
    primary alcohol or the corresponding aldehyde, $(H_3C)_2CHCH_2CHO$.
    d. $H_3CCH2CH2OH$; reduction of a carboxylic acid to yield a primary alcohol
    e.   Reversible dissociation of a carboxylic acid (a weak acid)

    f. $CO2$ + NADH; oxidative decarboxylation. The secondary alcohol group in isocitrate
    is oxidized to a ketone and loses a $CO2$ group.

4.   O...... HO-C-R
             || ||
          R-C-OH.......O

5. a. $Mg[H3C(CH2)14CO2]2$, a precipitate.
          Soaps precipitate in the presence of Ca^{+2} or Mg^{+2}, as found in hard water.
    b.   No reaction
          Detergents do not precipitate in these same conditions.
    c. $LiAlH4$ reduces a carboxylic acid to a primary alcohol.

6.   carboxylic acid > alcohol > aldehyde > alkane

7.   hydrogen bonding between molecules. The greater the hydrogen bonding between
molecules, the more energy it requires to vaporize the liquid and therefore the higher is the
boiling point.

8. $H_3C(CH_2)_7OH$; H_3CCO_2H

9. a.   3-Bromobenzoic acid
    b.   Methyl benzoate
    c.   Potassium 2,2-dichlorobutanoate

208

10.

a.

b.

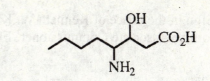

c.

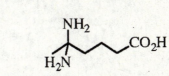

d.

e.

HO_2C CO_2H

QUICK QUIZ

18.1
(a) False   (b) False   (c) True   (d) False
(e) True   (f) False

18.2
(a) True   (b) True   (c) True   (d) False
(e) True   (f) True   (g) True   (h) True

18.3
(a) True   (b) False   (c) True   (d) True
(e) False   (f) False

18.4
(a) True   (b) True   (c) True   (d) True
(e) True   (f) True   (g) True   (h) True
(i) True   (j) True

18.5
(a) True   (b) True   (c) False   (d) False
(e) True   (f) True   (g) True   (h) False
(i) False   (j) False   (k) False

19

Carboxylic Anhydrides, Esters, and Amides

CHAPTER OBJECTIVES

After you have studied the chapter and worked the assigned exercises in the text and the study guide, you should be able to do the following.

1. Recognize the characteristic functional groups in carboxylic acids, the anhydrides derived from carboxylic acids and phosphoric acid, carboxylic esters, phosphate esters and amides.

2. Write out the structures for these compounds, in addition to those for cyclic esters (lactones) and cyclic amides (lactams).

3. Given the IUPAC name for an ester, amide or anhydride, write out the structural formula.

4. Given the structure for an ester, amide or anhydride, name the compound by either the IUPAC system or give its common name.

5. Indicate the monomers that are polymerized to form a) polyamides, b) polyesters and c) polycarbonates.

6. Draw the types of bonds that are produced on the formation of a) polyamides, b) polyesters and c) polycarbonates.

7. Contrast the physical properties of carboxylic acids with the comparable esters and amides and indicate how hydrogen bonding contributes to the differences in their boiling points and their solubility in water.

8. Write out the structural formulas for phosphoric acid and the monoester and the diester of phosphoric acid.

9. Write out structures for the anhydrides and esters of phosphoric acid and note the

similarities to carboxylic anhydrides and esters.

10. Write out a complete equation for the:
    a. reaction of a carboxylic acid and an alcohol to yield an ester.
    b. acidic (Fisher esterification) and basic (saponification) hydrolysis of an ester.
    c. reaction of an anhydride and an amine to yield an amide and a carboxylate salt.
    d. acidic and basic hydrolysis of an amide.
    e. hydrolysis of an anhydride to yield acids.
    f. reaction of an alcohol and an anhydride to an acid and an ester.
    g. reaction of an ester with ammonia to form an amide and alcohol.
    h. step-wise polymerization for polyamides, polyesters and polycarbonates.

11. Continue to write out summary sheets for the reactions associated with the new functional groups presented in this chapter. These should include:
    a. Carboxylic acids and the corresponding carboxylic esters, cyclic esters and anhydrides and the corresponding amides.
    b. Phosphoric acid and the corresponding phosphate esters and anhydrides.

12. Define the important terms and comparisons in this chapter and give specific examples where appropriate.

IMPORTANT TERMS AND COMPARISONS

Carbonyl Group and Carboxyl Group
Amines and Amides
Hydrophobic and Hydrophilic
Esters and Lactones (Cyclic Esters)
Fischer Esterification
Penicillins and β-Lactam Ring
Ester and Amide Hydrolysis Reactions
Aspirin (Acetylsalicylic Acid)
Barbiturates
Condensation and Step-Growth Polymers
Nylon 66, Kevlar, Dacron and Lexan

Esters and Anhydrides
Hydrogen Bonding
Anhydrides and Acyl Groups
Amides and Lactams (Cyclic Amides)
Saponification
Sunscreen and Sunblocks
Le Chatelier's Principle
Cyclooxygenase (COX) Enzymes
Phosphoric Anhydrides and Esters
Polyamides, Polyesters and Polycarbonates
Biodegradable

FOCUSED REVIEW

The two key functional groups in this chapter, from which all other derivatives (compounds) are derived, are:

1. Carboxylic acid
$$\underset{\displaystyle R-\overset{\displaystyle O}{\overset{\displaystyle \|}{C}}-OH}{}$$

2. Phosphoric acid
$$HO-\overset{\displaystyle O}{\overset{\displaystyle \|}{P}}-OH$$
$$\underset{\displaystyle OH}{|}$$

The derivatives, together with the so-called "parent" compound from which they are derived, are shown below.

	Carboxylic Acid	**Phosphoric Acid**

Acid

```
        O                          O
        ||                         ||
     R-C-OH                    HO-P-OH
                                   |
                                   OH
```

Anion(s)

```
        O           O  1-       O  2-        O  3-
        ||          ||          ||           ||
     R-C-O⁻      HO-P-O⁻      O-P-O        O-P-O
                    |           |            |
                    OH          OH           O
```

Ester(s)

```
        O           O           O            O
        ||          ||          ||           ||
     R-C-OR'     HO-P-OR'    R'O-P-OR'    R'O-P-OR'
                    |           |            |
                    OH          OH           OR'

                monoester     diester      triester
```

Anhydride

```
     O   O                   O   O
     ||  ||                  ||  ||
   R-C-O-C-R              HO-P-O-P-OH
                             |   |
                             OH  OH
```

Ester of the Anhydride

```
                             O   O
                             ||  ||
   _____                 ⁻O-P-O-P-OR
                             |   |
                             O⁻  O⁻


                            O    O    O
                            ||   ||   ||
                         ⁻O-P-O-P-O-P-OR
                            |    |    |
                            O⁻   O⁻   O⁻
```

These general examples of phosphate esters will become familiar in subsequent chapters, in the form of specific molecules that serve as metabolic intermediates (phosphoenolpyruvate, glucose phosphate and others) and in ATP (adenosine triphosphate) hydrolysis for energy production.

Amides

$$O$$
$$\|$$
$$R\text{-}C\text{-}NHR$$

In each of the derivatives, note exactly how the structure has been changed in going from the parent acid to the derivative compound. The similarities between the carboxylic acid and phosphoric acid should be clearly evident, along with the similarities in the analogous compounds derived from each parent. To become more familiar with these compounds and their chemistries, write out the complete equations for forming each derivative compound from the "parent" acid. Add these reactions to your summary sheets.

SELF-TEST QUESTIONS

<u>MULTIPLE CHOICE</u>. In the following exercises, select the correct answer from the choices listed.

1. The name of the molecule below is:

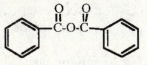

a.	Dibenzoic acid	c.	Benzoic anhydride
b.	Benzene anhydride	d.	Dibenzoic anhydride

2. Indicate which of the structures below is a lactam.

a.
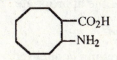

c.

b.

d.

3. The preferred strategy to prepare an amide is to react:

a. an amine with a carboxylic acid
b. an amine with an (carboxylic) acid anhydride
c. an amine with a carboxylic acid and then follow by reaction with strong base
d. an ester with an amine, followed by heat

4. The structural feature(s) that is common to all penicillins is (are):

a.	a carboxylic acid group	c.	a five-membered ring containing an S
b.	a four-membered lactone ring	d.	a four-membered lactam ring

5. The saponification of the compound, $H_3CCH_2CH_2\overset{\overset{O}{\|}}{C}$-$OCH_2CH_3$, in aqueous NaOH, yields:

   a. butanoic acid and ethanol
   b. ethanoic acid and butanol
   c. ethanol and sodium butanonate
   d. propanoic acid and ethanol

6. The basic hydrolysis of what amide produces the salt of pentanoic acid and dimethylamine.

a.	N, N-Dimethylpentanamide	c.	Dimethylpentanoic amide
b.	N-Dimethylpentanamide	d.	Pentanoic dimethylamide

7. Pentanolactone is an example of:

a.	an ester	c.	a cyclic ester
b.	an amide	d.	a cyclic amide

8. The ethyl group in N-ethylbutanamide is bonded directly on the

a.	carbonyl carbon	c.	carbon 2
b.	carbon 3	d.	the amide nitrogen

COMPLETION. Write the word, phrase or number in the blank or draw the appropriate structure in answering the questions.

1. Name the functional group or groups contained in each of the following condensed structural formulas. Some of the functional groups were presented in previous chapters.

   a.

   _____

b.

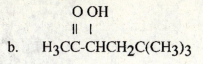

$$\underset{\substack{|| \quad |\\ O \quad OH}}{H_3CC-CHCH_2C(CH_3)_3}$$

_____

c.

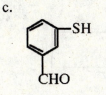

_____

d.

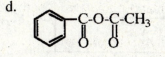

_____

e.

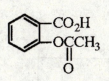

$$\underset{\substack{||\\O}}{(HO)_2POCH_2CH_3}$$

_____

f.

$$\underset{\substack{||\\O}}{(H_3C)_2CHOCCH_3}$$

_____

g.

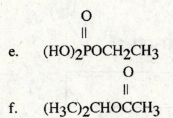

_____

h.

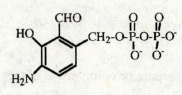

_____

2. Complete the following equations and name the type of reaction.

a. _____ + NaOH $\longrightarrow$ H$_3$CCO$_2^-$ Na$^+$ + CH$_3$CH$_2$OH

┌─────────────────────────┐
│   │
└─────────────────────────┘

215

b.

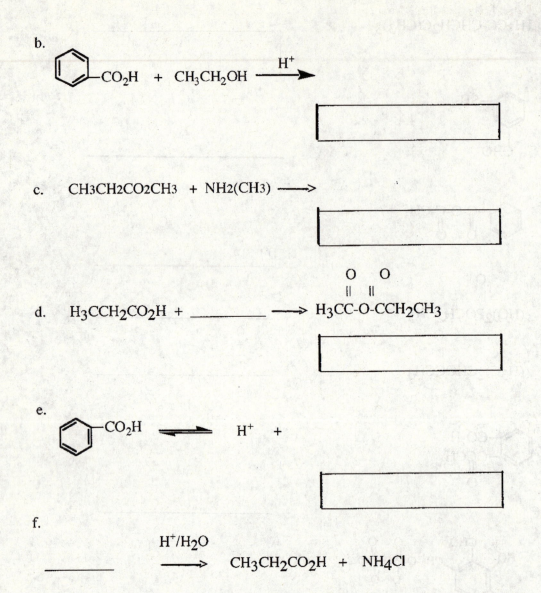

CO₂H + CH₃CH₂OH →(H⁺)

c. CH3CH2CO2CH3 + NH2(CH3) ──→

d. H₃CCH₂CO₂H + _____ ──→ H₃CC-O-CCH₂CH₃ (with two C=O groups: O O above the two C's)

e. ⬡-CO₂H ⇌ H⁺ +

f.

_____ ──→(H⁺/H₂O) CH₃CH₂CO₂H + NH₄Cl

3. Match the following type of polymer with its characteristic or commercial name.

a. Polycarbonate   i.   Kevlar
b. Polyamides   ii.   Made using phosgene
c. Polyester   iii. Biodegradable stitches
d. Polymerization of carboxylic acids   iv.   Dacron

4. Polyesters are examples of condensation polymers. To prepare these large molecules, each reactant molecule must have at least two functional groups. Draw out at least one repeating unit produced in the polyester formed in the reaction shown.

HOCH₂OH + HO₂CCH₂CH₂CH₂CO₂H ──→

216

5. The order of decreasing boiling points for an alkane, a carboxylic acid, an amide (with one N-H bond) and an alcohol, all of comparable molecular weight, is:

_____ > _____ > _____

6. Indicate the order of increasing solubility for the following compounds in water.

$CH_3CH_2CO_2H$, $CH_3CH_2CO_2NH_2$, $CH_3CH_2CH_2CH_2CH_3$, $CH_3CO_2CH_3$

_____ < _____ < _____ < _____

7. In the preparation of a polymer from two monomers that each has two functional groups, a part of the functional group is eliminated in the reaction. Indicate this elimination product in the reaction.

Polymer	Elimination product
a. Polyester	
b. Polyamide	
c. Polycarbonate	

8. Draw the structures for the following molecules.

   a. 2-Phenylpropanoic acid
   b. Phenyl-2-methylpropanoate
   c. Isopropy benzoate
   d. Heptanolactone
   e. N-Methyl-propanamide
   f. Ammonium pentanonate
   g. Methyl 3-phenylpentanonate

9. Aspirin is now known to _____ the synthesis of _____ by reacting with the enzyme, _____, which is often referred to by the acronym, COX.

QUICK QUIZ
   Indicate whether the statements are true or false. The number of the question refers to the section of the book that the question was taken from.

19.1 What Are Carboxylic Anhydrides, Esters, and Amides?
   (a) The functional group of an anhydride is two carbonyl groups bonded to the same oxygen atom.
   (b) The most polar bond of a carboxylic ester group is the C=O bond.
   (c) A lactone is a cyclic ester.
   (d) Formamide, the simplest amide, can be imagined to have been made with a primary amine.
   (e) A lactam is a cyclic amide.
   (f) An amide may participate in hydrogen bonding through both its carbonyl oxygen and its N- H groups.
   (g) A characteristic structural feature of the penicillins is a β-lactam.

(h) Esters and amides are polar molecules.

(i) An amide can be represented as a resonance hybrid of these two contributing structures.

19.2 How Do We Prepare Esters?

(a) Treatment of a carboxylic acid with an alcohol gives an ester.

(b) Fischer esterification results in substitution of the OH group of a carboxylic acid by the –OR group of an alcohol.

(c) Fischer esterification is an equilibrium reaction.

(d) The most common acid catalyst employed in Fischer esterification is acetic acid.

19.3 How Do We Prepare Amides?

(a) Treating a carboxylic acid with an amine at room temperature gives an amide and H_2O.

(b) Aspirin contains two functional groups; a carboxyl group and an amide.

(c) Treatment of an amine with an anhydride gives an amide and a carboxylic acid.

19.4 What Are the Characteristic Reactions of Anhydrides, Esters, and Amides?

(a) Hydrolysis of an anhydride gives two carboxylic acids.

(b) Acid-catalyzed hydrolysis of an ester gives a carboxylic acid and an alcohol,

(c) Acid-catalyzed hydrolysis of an ester is the reverse of Fischer esterification.

(d) Hydrolysis of an ester in aqueous NaOH gives an alcohol and the sodium salt of a carboxylic acid.

(e) Hydrolysis of an ester in aqueous NaOH requires one mole of base for each ester group.

(f) Hydrolysis of an ester in aqueous H_2SO_4 requires one mole of acid for each mole of ester group.

(g) Hydrolysis of a lactone can be accomplished by using either a catalytic amount of H_2SO_4 or a catalytic amount of NaOH.

(h) Of the three functional groups discussed in this chapter (anhydrides, esters, and amides), anhydrides are the most reactive toward hydrolysis, and esters are the least reactive.

(i) The sunscreen octyl p-methoxycinnamate (Chemical Connections 19C) contains two ester groups.

(j) The sunscreen Padimate A contains an ester group and an amide group.

(j) Treating an alcohol with an anhydride gives an ester and a carboxylic acid.

(k) Aspirin contains two functional groups; a carboxyl group and an ester group.

(l) Aspirin is more soluble in dilute aqueous NaOH than it is in pure water.

(m) Treating an amine with an anhydride gives two amides.

(n) Treating an ester with an amine gives an amide and an alcohol.

19.5 What Are Phosphoric Anhydrides and Phosphoric Esters?

(a) Phosphoric acid is a triprotic acid and forms mono-, di-, and triesters.

(b) In blood (pH 7.4), inorganic phosphate ion has a net charge of -2, and pyrophosphate ion has a net charge of -5.

(c) Phosphate diesters have a net negative charge at pH 7.4, the pH of blood.

19.6 What Is Step-Growth Polymerization?

218

(a) In the examples of step-growth polymerization presented in this chapter, each monomer contains two functional groups.

(b) Nylon-66 is so named because each monomer used in its synthesis has 6 carbon atoms; one is a six-carbon dicarboxylic acid, the other is a six-carbon diamine.

(c) Kevlar is a polyamide.

(d) Dacron polyester and Mylar are synthesized from the same dicarboxylic acid and a diol.

(e) The repeating functional group in a polycarbonate is a diester of carbonic acid.

(f) Glycolic acid and lactic acid, the monomers from which suturing-material, Lactomer, is synthesized, are constitutional isomers.

ANSWERS TO SELF-TEST QUESTIONS

MULTIPLE CHOICE

1. c	2. c	3. b	4. d
5. c	6. a	7. c	8. d

COMPLETION

1.
   a. lactone, ketone, lactam, aromatic ring
   b. ketone, secondary alcohol (note that this is not a carboxylic acid)
   c. thiol, aldehyde, aromatic ring
   d. anhydride, aromatic ring
   e. phosphate (mono) ester
   f. (carboxylic) ester
   g. carboxylic acid, carboxylic ester, aromatic ring
   h. phenol, aromatic amine, aldehyde, ester of a phosphate anhydride

2.
   a. $H_3CCO_2CH_2CH_3$; saponification of an ester

   b.

   $C\text{-}OCH_2CH_3$; Preparation of a (carboxylic) ester

   c. $H_3CCH_2CONH(CH_3)$; preparation of an amide.
   d. H_3CCO_2H; preparation of an acid anhydride
   e. Reversible dissociation of a carboxylic acid (a weak acid)

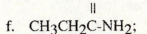

   f. $CH_3CH_2C\text{-}NH_2$;   Acid hydrolysis of an amide

219

3. a. ii   b. i   c. iv   d. iii

4. The following partial structure for the polymer shows one of the individual repeat units from the diol and the diacid underlined.

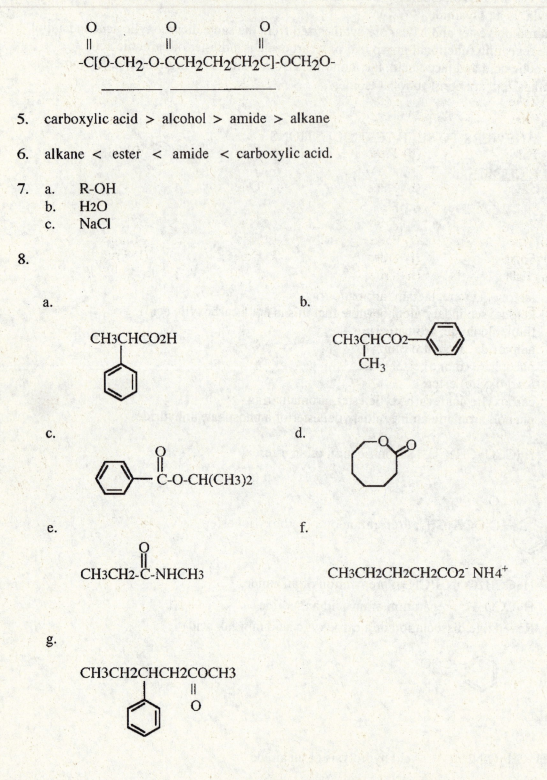

$$\underset{\qquad\qquad\quad}{\underset{\qquad}{-C[O\text{-}CH_2\text{-}O\text{-}CCH_2CH_2CH_2C]\text{-}OCH_2O-}}$$

with O above the C's

5. carboxylic acid > alcohol > amide > alkane

6. alkane < ester < amide < carboxylic acid.

7. a. R-OH
    b. H2O
    c. NaCl

8.

a.

CH3CHCO2H
(with phenyl ring below)

b.

CH3CHCO2— (phenyl ring)
$\quad\;\;$ CH$_3$

c.

(phenyl ring)—C-O-CH(CH3)2
(with O above C)

d.

(8-membered ring with O and =O)

e.

CH3CH2-C-NHCH3
(with O above C)

f.

CH3CH2CH2CH2CO2⁻ NH4⁺

g.

CH3CH2CHCH2COCH3
(phenyl ring below, O below CO)

9. inhibit, prostaglandins, cyclooxygenase

QUICK QUIZ

19.1

(a) True	(b) True	(c) True	(d) True
(e) True	(f) True	(g) True	(h) True
(i) True			

19.2

(a) False	(b) True	(c) True	(d) False

19.3

(a) False	(b) False	(c) True

19.4

(a) True	(b) True	(c) True	(d) True
(e) True	(f) False	(g) False	(h) False
(i) False	(j) True	(k) True	(l) True
(m) False	(n) True		

19.5

(a) True	(b) False	(c) True

19.6

(a) True	(b) True	(c) True	(d) True
(e) True	(f) False		

Burnt Sugar Icing Recipe
   2/3 cup granulated sugar
   1/3 cup boiling water
   3 cups powdered sugar
   2 tablespoons butter
      - H. Morgan
        You Can't Eat That!

20

Carbohydrates

CHAPTER OBJECTIVES

After you have studied the chapter and worked the assigned exercises in the text and the study guide, you should be able to do the following.

1. Define carbohydrates structurally and indicate a number of specific uses of carbohydrates in bacteria, plants and humans.

2. Draw the D form of the aldoses and ketoses that have three, four and five carbon atoms. Identify the stereocenter and the position that defines the sugar as being either the D or the L form.

3. Calculate the maximum number of stereoisomers (2^n rule) for aldoses and ketoses with 3, 4 and 5 carbon atoms. Draw the Fischer projection formulas for these and identify the enantiomeric pair of molecules and the diastereomers. Specify the D and L form of these sugars.

4. Indicate the physical and chemical properties that differ in enantiomers.

5. Describe in words the solution equilibrium for a monosaccharide such as β-D-glucose. Also draw the open chair and ring structures for the monosaccharides in the equilibrium. Identify the hemiacetal units in the Haworth formulas.

6. Write out the reactions for D-glucose undergoing (a) reduction, (b) oxidation and (c) phosphorylation. Note the atomic position at which the reaction takes place.

7.   List three monosaccharides and three disaccharides of importance. Name the products obtained from the hydrolysis of the specific disaccharides.

8.   Draw the Haworth structures for the three common disaccharides.

9.   Draw the Haworth structure for maltose, pointing out (a) the acetal unit, (b) the reducing ring and (c) the glycosidic bond.

10.   Describe the structural difference between amylopectin and amylose. Draw a minimum unit of this structure to show the difference.

11.   Describe the major structural difference between starch and cellulose. Explain the significance of the difference with regard to human metabolism.

12.   Characterize the bond types found in the hyaluronic acid and heparin. Explain why both polysaccharides are considered acidic.

13.   Define the important terms and comparisons in this chapter and give specific examples where appropriate.

IMPORTANT TERMS AND COMPARISONS

Carbohydrates
Aldoses and Ketoses
Chiral Molecule
Penultimate Carbon
Anomeric Carbon Atoms and Anomers
Cyclic Hemiacetals and Acetals
α- and β-D-Glucose
Alditols and Sorbitol
Reducing and Non-Reducing Sugar
L-Ascorbic Acid (Vitamin C)
Enantiomers and Diastereomers
Glucose, Dextrose, or Blood Sugar
Sucrose, Lactose and Maltose
Mutarotation
Saccharin and Artificial Sweetening Agents
Antigen and Antibody
Starch and Cellulose
α-1,4 and α-1,6 and β-1,6 Glycosidic bonds
Diabetes Mellitus
Acidic Polysaccharide
Hyaluronic Acid and Heparin

D- and L-Monosaccharides
Stereocenter
Fisher Projection
Haworth Projection
α- and β-Configuration of the
    Anomeric Carbon
Glycoside and Glycosidic Bond
Phosphoric Acid Esters
Chair Conformation for Cyclic Sugars
Aldonic Acids and Uronic Acids
Glycoside and Glucoside
Disaccharides and Polysaccharides
"Locked" Ring of a Disaccharide
Pyranose and Furanose Rings
A, B, AB and O Blood Groups
Oligo- and Polysaccharides
Amylose, Amylopectin and Glycogen
Hydrolysis and Condensation Reactions
Cellulose and β-Glucosidases
Glucose Oxidase
Enediol Intermediate

This chapter begins the formal coverage of **biochemistry** (concepts), including the nomenclature, structures, characteristics and reactions associated with these vital molecules. As will become evident, virtually all of these molecules, whether essential small molecules or biomacromolecules, have one, and in most cases, many stereocenters. As you study this and subsequent chapters, it is important to review Chapter 15 (Chirality:The Handedness of Molecules) and realize that this is an essential property of most molecules in living systems.

In addition, Chapter 17 (Aldehydes and Ketones) provides background for carbohydrates since, by definition, carbohydrates are polyhydroxyaldehydes, polyhydroxyketones or compounds that yield them after reaction with water (hydrolysis). Review these previous chapters as this new material is being assimilated.

Structures and Reactions

The cyclic forms of most monosaccharides contain a **hemiacetal** unit formed by the internal reaction of the alcohol group reacting with the carbon atom of the aldehyde group. Disaccharides contain an **acetal** unit and, if it includes a reducing sugar unit, it will also contain a hemiacetal unit. The formation and stability of these saccharide units is a major focus of this chapter. As introduced in Chapter 17, the reaction of an aldehyde or a ketone with an alcohol yields the formation of a hemiacetal.

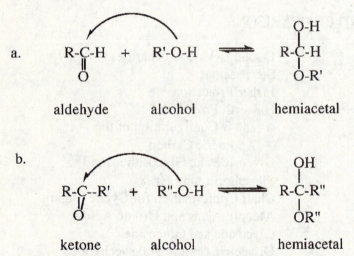

Recall that hemiacetals are unstable unless cyclic systems are formed as in the 5- or 6-membered furanose or pyranose rings, respectively.

For D-glucose, note the following:

(i)   the highest numbered stereocenter on the open-chain defines whether the monosaccharide is in the D- or L-series.

(ii)   the functional groups in the glucose

(iii) the equilibrium between the open-chain and the cyclic forms

(iv)   conversion of the open chain to the cyclic form produces an additional stereocenter

(v)   the position (i. e., the atom number) of the reacting groups in the open-chain form and the reactive group in the products

(vi)   the orientation of the groups on C-1 in the cyclic form

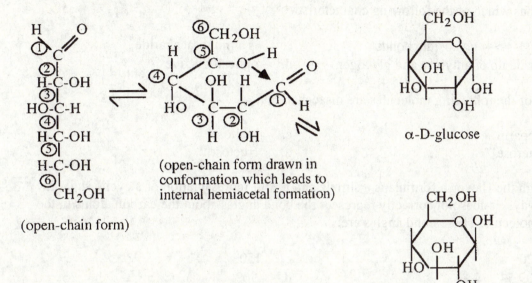

α-D-glucose

β-D-glucose

(open-chain form)

(open-chain form drawn in conformation which leads to internal hemiacetal formation)

Note that the anomeric carbon is at the C-1 position in the pyranose ring. As an exercise, draw the structures for D-fructose in an analogous reaction involving the formation of a five-membered ring containing a hemiacetal unit (formed from a ketone).

Two reactions to review are: (1) the oxidation of mono- or disaccharides that contain a hemiacetal unit and (ii) the hydrolysis reaction of the acetal bonds in polysaccharides that produces monosaccharides. The hydrolysis reaction is catalyzed by acids, or by enzymes in living systems.

SELF-TEST QUESTIONS

MULTIPLE CHOICE. Select the correct answer from the choices listed. In some cases, more than one answer is correct.

1.   Which of the following properties are associated with D-fructose?

    a.   optically active
    b.   undergoes mutarotation
    c.   disaccharide
    d.   is a ketose
    e.   is a reducing sugar
    f.   contains an α(1--->4) bond

225

2. Indicate which of the following properties is associated with maltose.

    a. rotates plane-polarized light
    b. contains a furanose ring
    c. contains a glycosidic bond
    d. is not a reducing sugar

3. Starch has which of the following characteristics?

    a. β (1--->4) glycosidic bonds
    b. made up of amylose and glycogen

    c. is a monosaccharide
    d. optically active

4. Which of the following molecules are disaccharides?

    a. heparin
    b. sucrose

    c. maltose
    d. fructose

5. Although the Haworth formula is a simple and useful representation of a cyclic sugar in many ways, it does not correctly represent the bond angles about the carbon atoms in the sugar molecule. These bond angles are:

    a. 120°
    b. 109.5°

    c. 150°
    d. 90°

6. The oxidation of D-glucose by glucose oxidase to produce D-gluconic acid occurs at which carbon atom?

    a. C-1
    b. C-2

    c. C-6
    d. C-4

7. The IUPAC name for linear form of D-ribose , an aldopentose, is

    a. 1, 2, 3, 4-Tetrahydroxypentanol
    c. 1, 2, 3, 4, 5-Pentahydroxypentanal

    b. 2, 3, 4, 5- Tetrahydroxypentanal
    d. 1-Tetrahydroxypentanal

8. D-glucose can be converted to L-ascorbic acid by some plants and animals, but not humans. L-Ascorbic acid (vitamin C) contains a number of functional groups, but does not contain a (an)

    a. lactone
    b. chiral carbon

    c. aldehyde
    d. alcohol functional group

9. Indicate which groups an acidic polysaccharides contains.

    a. phosphate
    b. sulfate

    c. acetate
    d. carboxyl

10. Indicate the number of stereocenters that the 5-carbon sugar, D-ribose, contains. Refer to Table 20.1 in the text.

a. 2   c. 4
b. 3   d. 5

11. The carbon in the primary alcohol group in D-glucose is carbon number:

a. 1   c. 4
b. 3   d. 6

12. The structure for a simple sugar is shown below. Indicate which characteristics it exhibits.

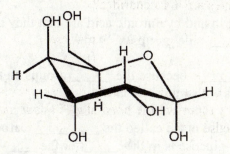

a. an aldopentose
b. a reducing sugar
c. shown in the α-configuration
d. is an acetal

COMPLETION. Complete the following statements with the appropriate word, phrase or number.

1. The process in which the α- and β-ring forms (hemiacetal) of a monosaccharide interconvert by way of the open chain structure is called _____.

2. Fructose is a _____ sugar, with a ketone group at the 2-carbon. The oxidation of a fructose occurs because in basic solution, it is in equilibrium with the _____ form. An _____, which contains a carbon-carbon double bond, serves as the key intermediate for this equilibrium and the oxidation.

3. The hydrolysis of sucrose produces _____ and _____.

4. The polysaccharide which contains the β(1-->4) glycosidic bond is _____.

5. The Haworth projection of glucose does not indicate that the glucose is actually in a _____ conformation, with the hydrogens or alcohol groups directed in either an _____ or an _____ projection.

6. The _____ (type of reaction) of a monosaccharide yields a product generally called an _____. This reaction specifically converts glucose to _____ or _____ (two alternative names).

227

7. Amylopectin differs from amylose in that it contains _____ that involve _____ glycosidic bonds.

8. Blood type _____ is the universal donor because it has _____ of the cell-surface bound carbohydrate _____ contained in A or B.

9. The acidic polysaccharide, _____, is naturally found in many human tissues. It is best known for its _____ activity in blood.

10. Lactose is an example of a _____ disaccharide because it contains a _____ unit. Therefore, it (will or will not) undergo mutarotation.

11. Two common features of heparin and hyaluronic acid are that they both contain _____ (rings) and _____ side groups.

12. Sugars are very soluble in _____ because the _____ groups on the sugar engage in _____ _____ interactions with water.

13. The number one carbon in glucose is the called the _____ carbon. It is this position that determines whether the D-glucose is in the _____- or the _____- form. These two forms of glucose are not enantiomers, but are _____.

14. _____ is the name of the disaccharide that contains only _____ glucose residues, in which a carbon on each glucose is bonded together by an _____ (a functional group) bond. This glycosidic bond is between carbon number _____ on one glucose and carbon number _____ on the other glucose. In the reaction of any two monosaccharides to yield a disaccharide, one molecule of _____ is also produced.

15. Humans cannot digest cellulose because they do not have the enzyme, _____, which cleaves a _____ bond in cellulose.

QUICK QUIZ
Indicate whether the statements are true or false. The number of the question refers to the section of the book that the question was taken from.

20.1 Carbohydrates: What Are Monosaccharides?
(a) Most carbohydrates contain only two types of functional groups: a carbonyl group and a hydroxyl group.
(b) The suffix -ose indicates that a compound is a carbohydrate.
(c) Fischer projection formulas show the configuration of all stereocenters in a monosaccharide.
(d) All carbon atoms in glucose are stereocenters.
(e) D- and L-glyceraldehyde are enantiomers.
(f) D-glucose and L-glucose are enantiomers.
(g) D-Fructose, a unit of sucrose, is the most abundant hexose in the biological world.
(h) Amino sugars contain a primary amino group in place of a carbonyl group.

228

20.2 What Are the Cyclic Structures of Monosaccharides?
   (a) Monosaccharides contain a carbonyl group and a hydroxyl group, and exist almost exclusively as cyclic hemiacetals.
   (b) Formation by a monosaccharide of a cyclic hemiacetal creates a new stereocenter.
   (c) Haworth projections show the configuration of all stereocenters, including that of anomeric carbons.
   (d) Only aldohexoses form cyclic hemiacetals.
   (e) α-D-glucose and β-D-glucose are mirror images of each other.
   (f) α-D-glucose and β-D-glucose differ in configuration only at carbon-5, the penultimate carbon.
   (g) L-Ascorbic acid (Vitamin C) contains a lactone (a cyclic ester) group.
   (h) In β-D-glucose, all groups on the six-membered cyclic hemiacetal ring are in equatorial positions.
   (i) Aqueous solutions of both α-D-glucose and β-D-glucose are dextrorotatory.
   (j) Mutarotation measures the change in the optical activity of an aqueous solution of a monosaccharide.

20.3 What Are the Characteristic Reactions of Monosaccharides?
   (a) A glycoside is a cyclic acetal.
   (b) The suffix -pyranoside indicates that a compound is a six-membered cyclic acetal of a monosaccharide.
   (c) The carbonyl group of D-glucose is reduced by $NaBH_4$ and by H_2/M to a primary hydroxyl group
   (d) Oxidation by a mild oxidizing agent converts the aldehyde group of D-glucose to a carboxyl group.
   (e) A reducing sugar is one that undergoes a reaction that reduces the number of oxygens in the molecule.
   (f) A reducing sugar is one that reduces an oxidizing agent.
   (g) D-Glucose and D-ribose are reducing sugars.
   (h) Treatment of D-glucose with Tollens' reagent gives D-gluconic acid.
   (i) The laboratory test for blood glucose levels is based on the oxidation of the aldehyde group of glucose to a carboxyl group.

20.4 What Are Disaccharides and Oligosaccharides?
   (a) A disaccharide contains two monosaccharide units joined by a glycosidic bond.
   (b) Sucrose is the most abundant disaccharide in the biological world.
   (c) Sucrose is a reducing sugar.
   (d) Lactose is a reducing sugar.
   (e) A, AB, B, and O blood types are determined by carbohydrates bound to plasma cell membranes.

20.5 What Are Polysaccharides?
   (a) Polysaccharides contain large numbers of monosaccharides linked by glycosidic bonds.
   (b) Complete hydrolysis of starch yields only D-glucose.
   (c) Cellulose is a linear polymer of D-glucose units joined by a 1,6-glycosidic bonds.

229

20.6 What Are Acidic Polysaccharides?

(a) Hyaluronic acid is present primarily in connective tissue.

(b) Hyaluronic acid is classified an acidic polysaccharide because one of the units from which it is synthesized is D-glucuronic acid, a compound that contains a carboxyl (-COOH) group.

(c) Heparin is a naturally occurring blood anticoagulant.

(d) Heparin contains $-COO^-$ groups and $-SO^{3-}$ groups.

ANSWERS TO SELF-TEST QUESTIONS

MULTIPLE CHOICE

1.	a, b, d, e	7.	b
2.	a, c	8.	c
3.	d	9.	b, d
4	b, c	10.	b
5.	b	11.	d
6.	a	12.	b, c

COMPLETION

1. mutarotation
2. ketose, aldose, enediol
3. glucose, fructose
4. cellulose
5. chair, axial, equatorial
6. reduction, alditol, D-glucitol, D-sorbitol
7. chain branching, (1-6)
8. 0, neither, antigens
9. heparin, anticoagulant
10. reducing, hemiacetal, will
11. pyranose, acidic
12. water, -OH, hydrogen bonding
13. anomeric, $\alpha-$, $\beta-$, diastereomers
14. Maltose, 2, acetal, 1, 4, water
15. β-glucosidases, $\beta(1-4)$

QUICK QUIZ

20.1

(a) True	(b) True	(c) True	(d) False
(e) True	(f) True	(g) False	(h) False

20.2

(a) True	(b) True	(c) True	(d) False
(e) False	(f) False	(g) True	(h) True
(i) True	(j) True		

20.3

(a) True	(b) True	(c) True	(d) True
(e) False	(f) True	(g) True	(h) True
(i) True			

20.4

(a) True	(b) True	(c) False	(d) True
(e) True			

20.5

(a) True	(b) True	(c) False

20.6

(a) True	(b) True	(c) True	(d) True

It takes a membrane to make sense
out of disorder in biology.
- Lewis Thomas
  The Lives of a Cell

21

Lipids

CHAPTER OBJECTIVES

Lipids are defined **operationally** - that is, they are defined by their properties and not their composition and structure. Molecules in a cell are collectively considered as lipids if they can be extracted from the cell extract with non-polar solvents. This classification procedure is not very rigorous and as a result a rather disparate melange of molecules, with very different compositions, structures and functions are enveloped within the definition. The focus of this chapter is therefore to define the (1) **COMPOSITION**, (2) **STRUCTURE**, (3) **PROPERTIES** and (4) **FUNCTION** of these cellular molecules. Keep this in mind as you review the material. As you will see, the functions range from providing the basic structure of all membranes, to highly specific hormones, some of which regulate metabolism, while others are very specific and define our sex.

After you have studied the chapter and worked the assigned exercises in the text and the study guide, you should be able to do the following.

1. Name the three major roles of lipids in human biochemistry.

2. List the four main groups of lipids and give a specific example of each.

3. Draw the general structure of a mono-, di- and triglyceride (all fats) which includes saturated and/or unsaturated fatty acids.

4. Explain how the degree of unsaturation in fatty acids influences their melting point.

5. Write out the general equations for the (1) hydrogenation of unsaturated fats and oils and for the (2) saponification of all fats or oils.

6. Define the lipid bilayer and describe the essential features of the fluid mosaic model of a membrane for a living cell.

7. Distinguish between transmembrane transport processes that take place by passive, facilitated and active transport mechanisms.

8. Draw a phosphoglyceride and a sphingolipid and indicate their main function.

9. Draw the structure for cholesterol and outline the relationship between cholesterol, progesterone, the adrenocorticoid hormones, and the sex hormones.

10. Characterize the four classes of lipoproteins in terms of composition, density and function.

11. Outline the interrelationship between the HDLs, LDLs, cholesterol and bile salts, and the important roles that each plays in the normal metabolism of cholesterol.

12. Outline the details of how bile salts and lipases work in concert to digest lipids.

13. Draw the structure of arachidonic acid and point out some common structural changes that occur on production of prostaglandins, leukotrienes and thromboxanes.

14. Discuss the function of cyclooxygenases, COX-1 and COX-2, in human physiology and the influence of nonsteroidal anti-inflammatory drugs (NSAIDs) on these critical enzymes.

15. Define the important terms and comparisons in this chapter and give specific examples where appropriate.

IMPORTANT TERMS AND COMPARISONS

Lipids
Saturated and Unsaturated Fatty Acids
Essential Fatty Acids
Hydrogenation
Soaps, Detergents, and Micelles
Phospholipids and Glycolipids
Membranes and Lipid Bilayer
Fluid Mosaic Model
Passive, Facilitated and Active Transport
Transmembrane Channels
Phosphatidylinositols (PI)
Sphingolipids and Sphingomyelin
Glycolipids-Cerebrosides and Gangliosides
Steroid and Cholesterol
HDL, LDL, VLDL and Chylomicrons
Bile Salts
LDL-Receptors
COX-1 and COX-2 Enzymes
Anabolic Steroids
Mineralcorticoids and Glucocorticoids
Sex Hormones - Testosterone and Estradiol
Progesterone and RU486
Lipases
Important Alcohols
    Glycerol
    Serine
    Choline
    Ethanolamine

Mono-, Di- and Triglycerides
Fats, Oils and Waxes
Polyunsaturated Oils
Saponification
Hydrophobic and Hydrophilic
Glycerophospholipids and Sphingolipids
    Phosphatidylcholine or Lecithin
Gap Junctions
Anion Transporters
Cephalin and Lecithin
Phosphatidylinositol 4, 5-Bisphosphate (PIP2)
Multiple Sclerosis and the Myelin Sheath
Lipid Storage Diseases and Tay-Sachs
Lipoproteins
    ApoB-100,LDL Receptors, Coated Pits and
    Endocytosis
HMG-CoA Reductase and Statin Drugs
Progesterone
Sex Hormones and Adrenocorticoid Hormones
Aldosterone and Cortisol
Oral Contraceptives
COX-1 and COX-2 Enzymes
Prostaglandins, Leukotrienes and Thromboxanes
Steroid and Nonsteroidal Anti-Inflammatory Drugs
    (NSAIDS)

233

Inositol
Sphingosine

SELF-TEST QUESTIONS

<u>MULTIPLE CHOICE</u>. In the following exercises, select the correct answer from the choices listed. In some cases, two or more choices will be correct.

1. Fats differ from waxes in that fats have:

   a. more unsaturation
   b. a glycerol backbone

   c. higher melting points
   d. longer fatty acids

2. Oleic acid is an unsaturated fatty acid containing:

   a. 12 carbons
   b. 14 carbons

   c. 16 carbons
   d. 18 carbons

3. Both stearic acid and linoleic acid contain 18 carbons. However, linoleic acid is unsaturated, while stearic acid is saturated. The melting point of stearic acid:

   a. is higher   b. is lower   c. is the same as linoleic acid
   d. relative to that for linoleic acid cannot be predicted

4. The most abundant steroid in the body is::

   a. cholesterol
   b. glucose

   c. Estrogen
   d. progesterone

5. Indicate which of the following structures represents cholesterol?

   a.

   b.

   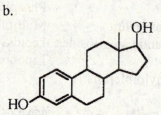

   c.

   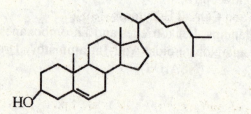

   d.

   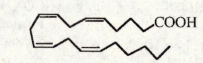

6. Prostaglandin group, PGE$_2$, possesses:

    a. ring closure at C-8 and C-12
    b. 20 carbon atoms
    c. a carbonyl group at C-9
    d. two olefinic (carbon-carbon) double bonds

7. Multiple sclerosis is thought to be associated with:

    a. degradation of the myelin sheath
    b. inhibited prostaglandin synthesis
    c. lipid storage
    d. a deficiency of enzyme COX-1

8. The main component in lipid bilayers are:

    a. complex lipids
    b. steroids
    c. prostaglandins
    d. fatty acids

9. Basic hydrolysis or saponification of triglycerides will produce:

    a. glycerol
    b. fatty acids
    c. salts of fatty acids
    d. bile salts

10. The 18-carbon backbone of sphingosine is found in:

    a. sphingomyelin
    b. myelin sheath
    c. phosphatidylcholine
    d. testosterone

11. Oils are:

    a. liquid triacylglycerides
    b. completely saturated
    c. high molecular weight fatty acids
    d. colored

12. Derivatives of what molecules are effective oral contraceptives?

    a. progesterone
    b. estradiol
    c. PGF$_{2\alpha}$
    d. cortisone

13. Low density lipoproteins (LDL) have which of the following characteristics:

    a. Contains more cholesterol than high density lipoproteins (HDL).
    b. Normal levels are soluble in the blood plasma
    c. Transports cholesterol to the cells
    d. Binds to LDL receptor molecules on the cell surface within the coated pits.

235

14. Anabolic steroids are

    a. powerful female sex hormones
    b. very effective detergents that aid digestion
    c. effective in the treatment of head aches
    d. used by athletes to increase muscle mass

15. The fluid mosaic model of the membrane is characterized by

    a. a lipid bilayer made up of complex lipids
    b. proteins embedded in the bilayer
    c. cholesterol molecules inserted in the bilayer
    d. all of these characteristics

16. Derivatives of this group of membrane lipids function as signaling molecules in chemical communication.

    a. cephalins   c. phosphatidylinositol
    b. ceramides   d. cerebrosides

17. The sex hormone that contains an aromatic ring is:

    a. progesterone   c. estrogen
    b. testosterone   d. aldosterone

18. Indicate which of the following characteristics bile salts exhibit.

    a. contain an aromatic ring
    b. derived from the oxidation of cholesterol
    c. provides a mechanism to eliminate excess cholesterol and its breakdown products
    d. contain either a glycine or taurine group

COMPLETION. Write the word, phrase or number in the blank space in answering the question.

1. Energy is stored in our bodies in the form of _____ because burning it produces twice as much energy as the burning of _____.

2. Aspirin has been found to inhibit _____ synthesis.

3. Both the physiologically potent prostaglandins and leukotrienes are derived from the 20-carbon, unsaturated fatty acid called _____ _____.

4. The enzymes involved in lipid metabolism are called _____.

5. One of the widely known diseases associated with lipid storage is _____.

6. _____ are negatively charged salts, which are powerful detergents used in the digestion of lipids. They are derived from the oxidation of _____.

236

7. The hydrophobic segment of a detergent is associated with the _____.

8. All unsaturated fatty acids exist as _____ at room temperature.

9. Rancidity is caused in part by the _____ of the fat or oil to aldehydes or by the production of _____ by the hydrolysis of triglycerides.

10. The fluid mosaic model of membranes proposes that _____ molecules are floating on and within the _____.

11. Fats are chemically the equivalent of an _____ (a functional group), composed of _____ and _____ with an even number of carbons. Waxes are similar, but are virtually always _____ (liquid or solid) because the alcohol component is a _____ _____ alcohol.

12. The common name for phosphatidylcholine is _____.

13. Leukotrienes are _____ of hormonal responses. Although similar in structure to prostaglandins, there is no _____ _____ in the structure of leukotrienes.

14. A faulty cholesterol transport system is often characterized by high levels of _____ and a relatively low level of _____. The disease, familial hypercholesterolemia, results from a deficiency of functional _____ _____ (two words).

15. Aldosterone is a _____ that regulates the concentrations of _____ and is synthesized in the _____ gland.

16. The target molecule for most statin drugs, that are used to inhibit cholesterol synthesis in the liver, is the enzyme, _____.

17. I was once told that the way to remember the steroid skeleton is to compare it to a house that contains "three rooms and a bath". In this analogy, the three rooms refer to the three _____ rings and the bath to the _____ ring.

18. Aspirin and other NSAIDs inhibit the enzymes, _____ and _____. NSAID is an acronym for _____ __-_____ _____ .

19. Under normal physiological conditions, COX-1 catalyzes the production of _____ from _____ _____ . On the other hand, when tissues are injured or damaged, the inflammatory response stimulates the production of _____ , which also produces prostaglandins.The best known _____ that are specific to COX-2 are (commercial names) _____ and _____ . Unfortunately, the use of at least one of these drugs has recently been correlated with the incidence of _____ _____ and _____ .

QUICK QUIZ

Indicate whether the statements are true or false. The number of the question refers to the section of the book that the question was taken from.

21.3 What Are Some of the Properties of Triglycerides?
   (a) All triglycerides contain saturated fatty acids.
   (b) All triglycerides contain unsaturated fatty acids.
   (c) The fatty acids that occur in triglycerides tend to have even numbers of carbon atoms.
   (d) Triglycerides that contain unsaturated fatty acids tend to be liquids at room temperature.
   (e) The base-catalyzed hydrolysis of triglycerides is called saponification.
   (f) Fatty acids tend to have many different kinds of functional groups in addition to the carboxyl group.

21.5 What Role Do Lipids Play in the Structure of Membranes?
   (a) Biological membranes consist entirely of lipids.
   (b) The lipids that occur in membranes have a hydrophobic portion and a hydrophilic portion.
   (c) The principal model for membrane structure is called the fluid mosaic model.
   (d) The higher the percentage of saturated fatty acids in a membrane, the more fluid it is.
   (e) Membrane lipids play an important role in the transport of substances into and out of the cell.

21.6 What Are Glycerophospholipids?
   (a) The structure of glycerophospholipids includes glycerol and phosphoric acid.
   (b) The structure of glycerophospholipids does not include fatty acids or nitrogen-containing compounds.
   (c) Glycerophospholipids play an important role in the structure of biological membranes.

21.8 What Are Glycolipids?
   (a) Glycolipids frequently occur in the brain.
   (b) Glycolipids tend not to contain sphingosine as part of their structure.
   (c) Ceramides frequently make up part of the structure of glycolipids.

21.9 What Are Steroids? The Structure of All Steroids is Based on a Characteristic Ring System.
   (a) Cholesterol is not found in membranes.
   (b) Lipoproteins of various densities are involved in transport of cholesterol in the blood.
   (c) Cholesterol levels in the blood are of little use as indicators of possible atherosclerosis.

ANSWERS TO SELF-TEST QUESTIONS

MULTIPLE CHOICE

1.	b	10.	a
2.	d	11.	a
3.	a	12.	a
4.	a	13.	a, b, c, d
5.	c	14.	d
6.	a, b, c, d	15.	d
7.	a	16.	c
8.	a	17.	c
9.	a, c	18.	b, c, d

238

COMPLETION

1. fat or triacylglycerides, carbohydrates
2. prostaglandin
3. arachidonic acid
4. lipases
5. any of the 5 listed in Box 21.E in the text, such as Gaucher's disease, Tay-Sachs and others
6. Bile salts, cholesterol
7. non-polar hydrocarbon chain
8. liquids
9. oxidation, fatty acids
10. protein, lipid bilayer
11. esters, glycerol, fatty acids, solid, long chain
12. lecithin
13. mediators, ring closure
14. LDL, HDLs, LDL receptors
15. mineralocorticoid, ions, adrenal
16. HMG-CoA reductase
17. 6-membered, 5-membered
18. COX-1, COX-2, nonsteroidal, anti-inflammatory, drugs
19. prostaglandins, arachidonic acid, COX-2, inhibitors, VIOXX, Celebrex, heart attacks, strokes

QUICK QUIZ

21.3

(a) False	(b) False	(c) True	(d) True
(e) True	(f) False		

21.5

(a) False	(b) True	(c) True	(d) False
(e) False			

21.6

(a) True	(b) False	(c) True

21.8

(a) True	(b) False	(c) True

21.9

(a) False	(b) True	(c) False

239

22

Proteins

CHAPTER OBJECTIVES

This chapter, similar to the previous chapter on lipids, concentrates on the (1) **COMPOSITION**, (2) **SEQUENCE**, (3) the **levels of STRUCTURE** and the (4) diverse **FUNCTIONS** of proteins. Proteins and/or enzymes are associated with, in some way, virtually all reactions in the cell and therefore will be of interest in all aspects of cellular regulation and metabolic processes.

After you have studied the chapter and worked the assigned exercises in the text and the study guide, you should be able to do the following.

1. List the eight major functions of proteins and give an example of an actual protein that is involved in each biological function.

2. Write out the structures for at least two amino acids which are representative of an amino acid with a (1) nonpolar, (2) polar, but neutral, (3) acidic and (4) basic R group. Also indicate the unique character of the amino acids glycine, proline and cysteine.

3. Demonstrate the amphiprotic character of amino acids by writing the reactions for the zwitterion with (1) an acid or (2) a base.

4. Write out the reversible reaction for the oxidation of cysteine and indicate how the production of cystine in a **protein** might stabilize the tertiary structure of proteins.

5. Draw the structure for a tripeptide, pointing out (1) the peptide bonds, (2) any stereocenters in the backbone, (3) the N- and C-terminal amino acid residues and (4) the R groups.

6. Name the four levels of structure in proteins and briefly describe what each is associated with.

7. Indicate the four forces that are involved in determining the final protein structure and draw a representation of each.

8. Characterize the secondary structures in an α-helix and a β-pleated sheet in terms of structure and the types of forces involved in each.

9. Describe the characteristics of a collagen triple helix and contrast it with the α-helix.

10. Discuss the structures for myoglobin and hemoglobin, the binding cooperativity exhibited by hemoglobin and contrast the hyperbolic and sigmoidal oxygen binding curves that characterize each one, respectively.

11. Indicate what changes occur in proteins by the addition of specific denaturation agents and list some practical applications of denaturation.

12. Define the important terms and comparisons in this chapter and give specific examples where appropriate.

IMPORTANT TERMS AND COMPARISONS

Protein and Peptide
Alpha Amino Acid and Amino Acid Residue
Hydrophobic and Hydrophilic Side Chains
Isoelectric Point, pI
Chains Amino Acid Names, 3-Letter- and 1-Letter Abbrev.
Cysteine and Cystine
Hypoglycemia
Fibrous and Globular Proteins
Di-, Tri-, Polypeptide and Proteins
(Nonenzymatic) Glycation
Diabetes and Aging
Primary, Secondary and Tertiary Structure
Salt Bridges and Hydrophobic Interactions
Intra- and Intermolecular Hydrogen Bonding
Conjugated Proteins and Prosthetic Group
Hemoglobin
Homo- and Heterozygote
Allosteric Proteins
Oxygen Binding Curves
Glycoproteins
Denaturing Agents and Denaturation
Proteome and Proteomics
Chaperone Proteins

Catalyst and Enzyme
Uncommon Amino Acids
Zwitterion
Nonpolar, Polar, Acidic & Basic Side
Amphiprotic Molecules
Thiol or Sulfhydryl (-S-H) Groups and
    a Disulfide Bond (-S-S-)
N-Terminal and C-Terminal Amino Acids
Peptide Backbone
AGE, Advanced Glycation End-product
Prion and Pathological Conditions
Quaternary Structure
α-Helix, β-Pleated Sheet and Random Coil
Triple Helix Structure of Collagen
Tropocollagen and Collagen
Sickle Cell Anemia
Alpha Chains in Globular Hemoglobin
Positive Cooperativity
Hyperbolic or Sigmoidal Shaped Curves
O-Linked and N-Linked Saccharides
Integral Membrane Proteins
Protein Microarrays
Protein Functions
    Structure
    Catalysis
    Movement
    Transport
    Hormones
    Protection
    Storage
    Regulation

241

SELF-TEST QUESTIONS

<u>MULTIPLE CHOICE</u>. In the following exercises, select the correct answer from the choices listed. In some cases, two or more choices will be correct.

1. The number of tetrapeptides that can be made from any of the 20 amino acids is:

    a. 1.6×10^5 c. 4
    b. 20   d. 80

2. Glycine is an unique amino acid because it:

    a. has no enantiomer
    b. has a sulfur containing R group
    c. cannot form a peptide bond
    d. is an essential amino acid

3. The α-helix is a common form of:

    a. primary structure   c. tertiary structure
    b. secondary structure   d. a denatured region

4. Which of the following is a transport protein?

    a. hemoglobin   c. collagen
    b. pepsin   d. oxytocin

5. A zwitterion has which of the following characteristics?

    a. a high melting point
    b. no net charge
    c. soluble in water
    d. all of above

6. Arginine is a basic amino acid, while alanine is nonpolar. The isoelectric point for arginine is:

    a. higher than for alanine   c. about the same as alanine
    b. lower than for alanine   d. cannot predict

7. Hemoglobin is an oxygen carrying protein made up of two α-chains and two β-chains. Hemoglobin has:

    a. 2 subunits   c. no subunits
    b. 1 subunit   d. 4 subunits

8. Which of the following R groups associated with an amino acid can take part in hydrogen bonding?

242

a. -CH$_2$SCH$_3$ (methionine)   c. -CH$_2$CO$_2$H (aspartic acid)

b. -CH$_2$OH (serine)   d. -CH$_2$CH$_2$CH$_2$CH$_2$NH$_3^+$ (lysine)

9. Hydrophobic interactions occur between the R groups with:

a. acidic character   c. basic character

b. nonpolar character   d. all of above

10. The number of amino acid residues in the cyclic nonapeptide, vasopressin, is:

a. 7   c. 19

b. 9   d. 90

11. An example of a globular protein that is a hormone is

a. hemoglobin   c. collagen

b. keratin   d. insulin

12. The secondary structure in an α-helix is stabilized (held together) by

a. intramolecular hydrogen bonds   c. electrostatic bonds

b. hydrophobic interactions   d. intermolecular hydrogen bonds

13. An example of a prosthetic group is

a. methionine   c. an amide group

b. the heme group   d. any functional group

14. Muscle proteins that are involved in movement are:

a actin and myosin   c. collagen and protocollagen

b. heme and hemoglobin   d. fibrinogen and platelets

15. Two amino acid residues **in a protein** that can take part in electrostatic interactions are:

a. alanine and isoleucine   c. methionine and cysteine

b. histidine and lysine   d. glutamic acid and phenylalanine

16. The charge on the tripeptide, leu-arg-lys, at pH 7, is:

a. +1   c. +3

b. +2   d. zero

17. The ingestion of heavy metals, such as lead (Pb^{+2}) or mercury (Hg^{+2}) can be poisonous because they denature proteins by binding to:

a. methionine   c. arginine

243

b.   cysteine   d.   tyrosine

18.   Indicate which of the following proteins are fibrous proteins.

a.   myoglobin   c.   keratin
b.   collagen   d.   hemoglobin

19.   The complete complement of proteins that is expressed in a cell is called the:

a.   proteome   c.   genome
b.   amphiprotic groups   d.   polypeptides

20.   Indicate the protein that does not have a quaternary structure.

a.   myoglobin   c.   hemoglobin
b.   collagen   d.   phenylalanine

21.   Indicate which of the following characteristics hemoglobin exhibits.

a.   a sigmoidal oxygen binding curve   c.   a quaternary structure
b.   an oxygen storage protein   d.   oxygen binding to hemoglobin shows
                                                 positive cooperativity

COMPLETION. Write the word, phrase or number in the blank space in answering the question.

1.   In the tripeptide drawn below, circle and clearly label the following units: peptide
linkages;
     C- and N-terminal end; R groups.

$$
\begin{array}{c}
\text{H O}\quad\text{H O}\quad\text{H} \\
|\ ||\quad\ |\ ||\quad\ | \\
{}^{+}\text{H3N-C-C-N-C-C-N-C-COO}^{-} \\
|\quad\ |\quad|\quad\ |\ | \\
\text{H2C}\quad\text{H}\quad\text{CH3 H CH2} \\
|\qquad\qquad\qquad\ | \\
\text{H2C}\qquad\qquad\ \text{CH2} \\
|\qquad\qquad\qquad\ | \\
\text{COO}^{-}\qquad\qquad\text{CH2} \\
\qquad\qquad\qquad\ | \\
\qquad\qquad\qquad\text{CH2} \\
\qquad\qquad\qquad\ | \\
\qquad\qquad\qquad\text{NH3}^{+}
\end{array}
$$

2.   Tropocollagen is the _____ form of collagen and is made up of _____

_____ units. Insoluble collagen results from covalent _____ of two _____ residues on adjacent chains in the helix.

3. A covalent bond in the _____ _____ (protein backbone), is the force which is associated with the _____ structure of a protein.

4. Agents that can reversibly denature a single subunit protein do this by breaking forces involved in the _____ and _____ (levels of) structure.

5. Sickle cell trait occurs in people who have _____ gene(s) programmed to produce the protein in sickle cell hemoglobin.

6. The α-helix is stabilized by intramolecular hydrogen bonding between _____ amino acid residues along the peptide chain, while the hydrogen bonding in forming a β-pleated sheet structure is stabilized by _____ or _____ hydrogen bonding between amino acid residues. The interaction in both cases, is between the _____ and the _____ groups in the peptide _____.

7. N-Linked saccharides have the saccharide linked directly to _____, while O-linked saccahrides may be linked to either _____, _____ or _____.

8. Complete the following reactions:

a.
$$\overset{\displaystyle \overset{H}{|}}{H_3N-C-CO_2^-} + H^+ \longrightarrow \qquad \underset{\displaystyle |}{\underset{\displaystyle R}{}}$$

_____

b.
$$2 \; \overset{\displaystyle \overset{H}{|}}{\underset{\displaystyle \underset{CH_2}{\underset{|}{SH}}}{H_3N-C-CO_2^-}} \; \begin{smallmatrix} [O] \\ \longrightarrow \\ \longleftarrow \\ [H] \end{smallmatrix}$$

_____

c. tetrapeptide + HCl $\overset{heat}{\longrightarrow}$ _____

d.

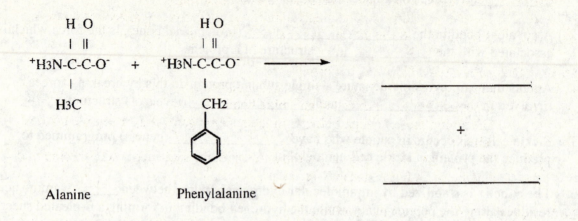

Alanine   Phenylalanine

9. _____ structure occurs only in proteins that have _____ subunits, such as in hemoglobin.

10. The charge on hydrophobic amino acid, alanine, at very low pH is _____, while at high pH it is _____.

11. _____ cooperativity, which is found in hemoglobin, means that after the _____ oxygen binds, the next oxygen binds much _____.

QUICK QUIZ
   Indicate whether the statements are true or false. The number of the question refers to the section of the book that the question was taken from.

22.2 What Are Amino Acids?
   (a)   Nineteen of the common 20 amino acids have the same general formula where only the R group differs.
   (b)   The amino acids that make up proteins are α-amino acids
   (c)   The amino acids found in your proteins have the D-configuration
   (d)   The three letter abbreviation for glutamine is Glu
   (e)   The one-letter abbreviation for lysine is L
   (f)   Serine is a polar, neutral amino acid.
   (g)   Histidine is an acidic amino acid because its side chain has a pKa of 6.0.

22.3 What Are Zwitterions?
   (a)   A zwitterion has a positive charge on one atom and a negative charge on another
   (b)   Amino acids are zwitterions only in solution.
   (c)   An acidic amino acid can only be a zwitterion at its isoelectric point.
   (d)   Most of the 20 common amino acids have isoelectric points above 7.
   (e)   An amino acid could make a buffer for at least two different pH ranges.

22.4 What Determines the Characteristics of Amino Acids?
   (a)   The unique characteristic of an amino acid is its α-carboxyl group.
   (b)   The unique characteristic of glutamic acid is its α-carboxyl group.

(c)   Cysteine has an oxidized and a reduced form.

(d)   All of the aromatic amino acids act as neurotransmitters

(e)   Tyrosine, tryptophan, and phenylalanine are precursors of neurotransmitters.

(f)   Histidine's side chain is always charged at physiological pH

22.6 How Do Amino Acids Combine to Form Proteins?

(a)   Amino acids are linked together via peptide bonds.

(b)   A peptide bond is also called an amide bond.

(c)   The amino group that participates in the reaction generating a peptide bond is called the N-terminus.

(d)   The sequence KLA indicates a tripeptide with the sequence lysine-leucine-alanine, with lysine as the N-terminus.

22.8 What Is the Primary Structure of a Protein?

(a)   Using three different amino acids, one could make 6 different tripeptide.

(b)   Using the 20 common amino acids, there are 20^{50} possible sequences of a protein with 50 amino acids

(c)   The decision to name peptides from N-terminus to C-terminus was completely arbitrary.

(d)   Amino acid substitutions are always detrimental to a protein.

(e)   A single amino acid substitution in the β–chain of sickle cell hemoglobin accounts for the effects of the disease.

22.9 What Is the Secondary Structure of a Protein?

(a)   The α-helix is an example of secondary structure

(b)   The α-helix can be formed between two polypeptide chains.

(c)   Secondary structures are formed by hydrogen bonding between groups on the peptide backbone.

(d)   The β-pleated sheet can be formed between groups in a single polypeptide or between different polypeptides

(e)   Prions are small proteins found only in sheep and cattle that can infect humans.

(f)   There is an inheritable component to susceptibility to prion diseases.

22.10 What Is the Tertiary Structure of a Protein?

(a)   The side chain of serine could make a hydrogen bond with a carbonyl oxygen that is part of the peptide backbone.

(b)   When the side chain of glutamic acid interacts with the side chain of lysine, the result is called a hydrophobic interaction.

(c)   Two side chains with the same net charge can be linked together via metal ion coordination.

(d)   Hydrophobic interactions are the weakest of the interactions that are part of tertiary structure.

(e)   The study of how many genes can be found in a given organism is called proteomics.

(f)   The complement of proteins being produced in a given organism is called the proteome.

22.11 What Is the Quaternary Structure of a Protein?

(a)   All proteins have quaternary structure.

(b)   In quaternary structure, subunits are held together by disulfide bonds.

(c)   All proteins that have quaternary structure also exhibit allosteric behavior.

(d)   The effect of binding one ligand to a subunit making it easier to bind the ligand to a second subunit is called positive cooperativity.

(e)   Hemoglobin is a conjugated protein.

(f)   Heme is a prosthetic group in both hemoglobin and myoglobin.

(g)   Myoglobin has a higher affinity for oxygen than hemoglobin.

(h)   The binding curve of hemoglobin for oxygen is hyperbolic.

ANSWERS TO SELF-TEST QUESTIONS

MULTIPLE CHOICE

1. a
2. a
3. b
4. a
5. d
6. a
7. d
8. b, c, d
9. b
10. b
11. d

12. a
13. b
14. a
15. b
16. b
17. b
18. b, c
19. a
20. a
21. a, c, d

COMPLETION

1.

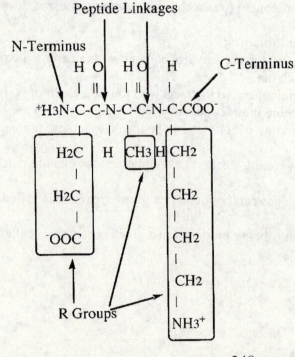

2. soluble; triple helical; crosslinking; lysine
3. peptide (bonds) linkage; primary
4. secondary; tertiary
5. one
6. nonadjacent, intramolecular, intermolecular, $\backslash$C=O, $/$H-N , backbone
                                                    $/$ $\backslash$

7. asparagine, serine, threonine, hydroxylysine
8.

a.
$$\overset{\quad H}{\underset{\quad R}{H_3\overset{+}{N}-C-CO_2H}}$$

b.
$$\underset{\quad H \qquad\qquad\qquad\quad H}{\overset{CO_2^- \qquad\qquad\qquad CO_2^-}{H_3\overset{+}{N}-C-CH_2-S-S-CH_2-C-NH_3^+}}$$

cystine, containing a disulfide bond

c. an acid hydrolysis reaction to yield 4 amino acids

d. Two different dipeptides are produced with the same composition, but with different
    sequence. In addition, the "homodipeptides", ala-ala and phe-phe, should also be
    produced, making a total of four.

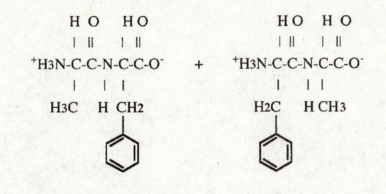

9. Quaternary, multiple (2 or more)
10. +1, -1
11. Positive, first, easier

QUICK QUIZ

22.2

(a) True	(b) True	(c) False	(d) False
(e) False	(f) True	(g) False	

22.3

(a) True	(b) False	(c) False	(d) False
(e) True			

22.4

(a) False	(b) False	(c) True	(d) False
(e) True	(f) False		

22.6

(a) True	(b) True	(c) False	(d) True

22.8

(a) False	(b) True	(c) False	(d) False
(e) True			

22.9

(a) True	(b) False	(c) True	(d) True
(e) False	(f) True		

22.10

(a) True	(b) False	(c) True	(d) True
(e) False	(f) True		

22.11

(a) False	(b) False	(c) False	(d) True
(e) True	(f) True	(g) True	(h) False

Were it not equipped with catalysts, every
living unit would be a static system
   - F.G. Hopkins
    Science, 78, 219 (1933)

23

Enzymes

CHAPTER OBJECTIVES

After you have studied the chapter and worked the assigned exercises in the text and the study guide, you should be able to do the following.

1. Explain the statement that enzymes increase the rate of the reaction, but do not change the position of the equilibrium.

2. List the six major categories of enzymes and indicate the general type of reaction that each category of enzyme catalyzes.

3. Distinguish between the terms, enzyme, cofactor, coenzyme, apoenzyme, proenzyme, isoenzyme, active site and regulatory site.

4. Describe, in both words and in figures, the effect the following have on the rate of an enzyme-catalyzed reaction:
    > enzyme concentration
    > substrate concentration
    > temperature
    > pH
    > enzyme inhibitors

5. Outline and compare the proposals in the (1) lock and key model and (2) induced fit model to explain enzymatic action and specificity.

6. Describe how competitive and a non-competitive reversible inhibitors produce their effect.

7. Define active site and list factors associated with it that are responsible for its specificity toward molecular substrates.

8. Explain how the activity of an allosteric enzyme can be modulated in a positive or negative way by regulator interactions at the regulatory site. Relate this effect to the feedback control mechanisms found in multiple-step reaction schemes.

9. Distinguish between the terms enzymes, ribozymes and abzymes.

10. Define the important terms and comparisons in this chapter and give specific examples where appropriate.

IMPORTANT TERMS AND COMPARISONS

Enzyme and Catalyst
Substrate and Competitive Inhibitor
-ase
Substrate and Active Site
Enzyme Specificity and Activity
Transition State
Proenzyme and Zymogens
Reversible and Irreversible Inhibition
Inhibitors
Oxidoreductase
Transferase
Hydrolase
Lyase
Isomerase
Ligase
Active Site and Regulator Site
Regulatory or Allosteric Enzymes
Regulator and Regulator Site
Isozymes or Isoenzymes
Immunogen
Feedback Control of Enzyme Activity
Kinases and Phosphatases

Ribozymes
Cofactor and Coenzyme
Apoenzyme and Holoenzyme
Enzyme-Substrate Complex
Active Site
Transition State Analogs
Activation and Inhibition
Competitive and Non-Competitive

pH Profile for Enzymatic Activity
Lock-and-Key and Induced-Fit Models
Irreversible Inhibitor
Sulfa Drugs and p-Aminobenzoic Acid
Le Chatelier's Principle
Allosterism and Allosteric Enzymes
Positive & Negative Modulation of
    Regulatory Enzyme Activity
T and R Form of Allosteric Enzymes
Proteases
Catalytic Antibodies and Abzymes
Protein (Enzyme) Modification

SELF-TEST QUESTIONS

MULTIPLE CHOICE. In the following exercises, select the correct answer from the choices listed. In some cases, two or more choices will be correct.

1. A molecule that is structurally similar to the substrate for an enzyme will probably be a:

   a. competitive inhibitor
   b. cofactor
   c. regulator
   d. noncompetitive inhibitor

2. The site on an allosteric enzyme that is directly involved in modulation of its activity is called

   a. the prosthetic group
   b. regulatory site
   c. active site
   d. target site

252

3. The multistep sequence of reactions shown below uses a regulatory mechanism to control the amount of D produced. In these general schemes, the last product, D, in the reaction sequence is involved in a feedback mechanism of inhibition. Molecule D in the sequence usually

a. inhibits E_1

c. inhibits E_3

b. inhibits E_2

d. inhibits all enzymes

$$E_1 \qquad E_2 \qquad E_3$$
$$A \text{ -----> } B \text{ -----> } C \text{ -----> } D$$

4. Competitive inhibition can be overcome by:

a. increasing substrate concentration
b. increasing pH
c. decreasing temperature
d. all of above

5. Most enzymatic reactions are influences by pH because pH changes can:

a. completely hydrolyze the protein
b. produce protonation or deprotonation of essential amino acid residues in the active site
c. change its primary structure
d. effect the optical activity of the protein

6. A significant increase in the temperature (as when a patient has a very high fever) of an enzymatic reaction may reduce the catalytic rate because the:

a. enzyme acts on itself as an inhibitor
b. protein is partially or completely denatured
c. protein undergoes hydrolysis
d. heat acts as a competitive inhibitor

7. Consider the possible shapes that a small molecule might have to act as a competitive inhibitor to the reaction shown below:

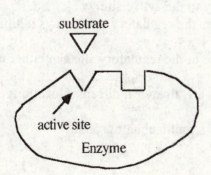

253

Indicate which of the following molecules (molecular shapes) could potentially act as a competitive inhibitor.

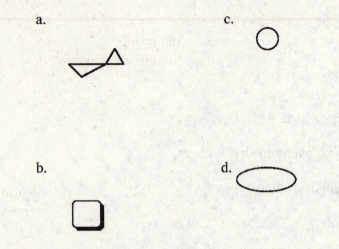

a.

c.

b.

d.

8. The apoprotein corresponds to which part of an enzyme?

a. coenzyme
b. protein portion

c. cofactor
d. part cleaved off in a proenzyme

9. The site on the enzyme at which the substrate interacts is called the

a. regulatory site
b. modulator site

c. active site
d. allosteric site

10. A kinase is an enzyme that acts on another enzyme, A, to change its activity. A kinase modifies enzyme A by

a. cleaving enzyme A to produce a smaller active enzyme.
b. adding a missing coenzyme in the reaction.
c. adding a phosphate group on the enzyme.
d. adding a glycosyl group on the enzyme.

11. Indicate which of the following characteristics an allosteric enzyme exhibits.

a. A regulator molecule can bind strongly to the active site.
b. The binding of a regulatory molecule to the regulatory site always inhibits enzymatic activity.
c. The binding of the regulatory molecule to the regulatory site can either increase or decrease the enzymatic activity.
d. The active site and the regulatory sites are always in different locations and cannot be identical.
e. In most cases, the allosteric enzyme is a multisubunit protein.

12. Characteristics of a proenzyme are:

a. they are made in the cell in an inactive form.
b. they do not contain all the amino acid residues for a complete and active enzyme.
c. many proenzymes are usually cleaved (cut) to produce a smaller protein that becomes the active enzyme.
d. they do not have the correct prosthetic group for activity.

13. The protein modification that is most common in the activation of enzymes is

a. methylation
b. phosphorylation
c. addition of an aromatic group
d. addition of an ester group

14. The enzyme, prostaglandin enderoperoxide synthase (PGHS), (Box 22F) is <u>unusual</u> in that it

a. has two active sites and catalyzes two different reactions
b. has two regulatory sites and no active site
c. can be inhibited by common over-the-counter-drugs
d. is primarily composed of a polysaccharide

15. The small molecule, urea, is used to denature proteins. Its structure is $H_2N-C-NH_2$
Its ability to denature proteins is primarily because $\overset{\displaystyle \|}{\underset{\displaystyle O}{}}$

a. it cleaves the peptide backbone
b. it breaks up essential hydrogen bonding in the secondary and tertiary structure
c. it breaks up hydrophobic interactions in the secondary and tertiary structure
d. it reacts with the hydroxyl group on glycine residues

16. Isoenzymes are enzymes that

a. occur in different forms in different tissues and catalyze the same reaction.
b. are always composed of only one subunit.
c. must be modified by a kinase to become active.
d. are unaffected by extreme temperatures.
e. are composed of RNA.

<u>COMPLETION</u>. Write the word, phrase or number in the blank space in answering the question.

1. Classify each of the following enzymes into one of the six major classes.

Enzyme	**Class**
a. Acid Phosphatase	
b. Alcohol Dehydrogenase	
c. DNA (Topo)isomerase	
d. Pepsin	

    e.   Tyrosine-tRNA Synthetase
    f.   Pyruvate Kinase
    g.   Acetylcholinesterase
    h.   Aspartate Amino Transferase

2.   Lipases are enzymes that are (<u>more or less</u>) specific than ACE.

3.   Of the two main models used to explain enzymatic action, the _____ model is regarded as more flexible than the _____ model.

4.   A number of enzymes require a non-protein component called a _____ or _____ to exhibit enzymatic activity.

5.   Trypsinogen is an example of a _____ that can only be _____ by removal of a portion of the protein.

6.   If the enzyme concentration is tripled, the reaction rate will _____. (Refer to Figure 23.3 in text).

7.   Enzymes change the _____ of the reaction, but do not change the _____ concentrations in the reaction.

8.   Label the following plot to show a pH profile for an enzyme in which the greatest activity occurs at pH 4.5 and decreases to nearly zero at pH 7.0 and 2.0. Make the plot with enzyme rate on the vertical axis and pH on the horizonal axis.

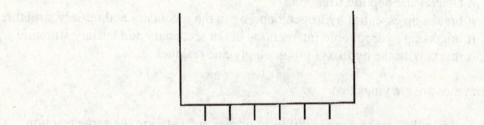

9.   Draw the substrate, enzyme and enzyme-substrate (E-S) complex as they may be viewed in the induced fit model.

_____ + _____ _____
Substrate   +   Enzyme   (E-S) complex

10.   Although cofactors can be either _____ or _____, coenzymes are always _____ and are synthesized from a vitamin.

11.   The five amino acid residues that are found most often in the active site of an enzyme are _____, _____, _____, _____ and _____ . Of these amino acids, 2 are _____ amino acids, 2 are _____ amino acids and the last one contains a _____ group.

QUICK QUIZ

Indicate whether the statements are true or false. The number of the question refers to the section of the book that the questions were taken from.

23.1 What Are Enzymes?
- (a)   All enzymes are globular proteins.
- (b)   Some enzymes are made out of DNA.
- (c)   Enzymes are biological catalysts.
- (d)   Enzymes exhibit substrate specificity.
- (e)   Trypsin is an enzyme that catalyzes the cleavage of peptide bonds.
- (f)   Trypsin makes the free energy of activation of the reaction more favorable.
- (g)   Lipases catalyzes the cleavage of fatty acids off of triacylglycerols.

23.2 How are Enzymes Named and Classified?
- (a)   Non-protein parts of enzymes are called cofactors.
- (b)   The polypeptide portion of an enzyme is called an apoenzyme.
- (c)   Heavy metal cofactors are called coenzymes.
- (d)   The molecules that the enzyme binds to and converts to products are called substrates.
- (e)   When an inhibitor binds to the active site, it is called a non-competitive inhibitor.
- (f)   In a reaction with only one substrate, a competitive inhibitor and the substrate cannot bind at the same time.

23.4 What Factors Affect Enzyme Activity?
- (a)   If you double the amount of enzyme, the rate of the enzyme-catalyzed reaction doubles.
- (b)   If you double the amount of substrate, the rate of the enzyme-catalyzed reaction doubles.
- (c)   If you double the temperature of an enzyme catalyzed reaction, the reaction rate will double.
- (d)   An enzyme has a characteristic profile of rate versus pH.

23.6 How Are Enzymes Regulated?
- (a)   Feedback control means that the product of a reaction inhibits the enzyme that catalyzed its production.
- (b)   An enzyme that is produced initially in an inactive form is called a zymogen.
- (c)   The zymogen form of an enzyme and the active form are easily interconverted.
- (d)   Allosteric enzymes often have multiple subunits.
- (e)   A substance that binds to an allosteric enzyme at a location other than the active site is called a regulator.
- (f)   Glycogen phosphorylase is an allosteric enzyme
- (g)   Glycogen phosphorylase is regulated by phosphorylation.
- (h)   Kinases and phosphatases are often used to modify enzymes.
- (i)   Enzymes that have multiple forms in different tissues are called isozymes.
- (j)   In a given organism, there are 8 isozymes of lactate dehydrogenase.
- (k)   Isozymes of lactate dehydrogenase differ in their electrophoretic mobility.

257

MULTIPLE CHOICE

1. a
2. b
3. a
4. a
5. b
6. b
7. a
8. b

9. c
10. c
11. c, d, e
12. a, c
13. b
14. a
15. b
16. a

COMPLETION

1. a. Hydrolase
    b. Oxidoreductase
    c. Isomerase
    d. Hydrolase

    e. Ligase
    f. Transferase (specific for phosphates)
    g. Hydrolase
    h. Transferase

2. less
3. induced fit; lock and key
4. cofactor, coenzyme
5. proenzyme, activated
6. triple
7. rate, (thermodynamic) equilibrium

8.

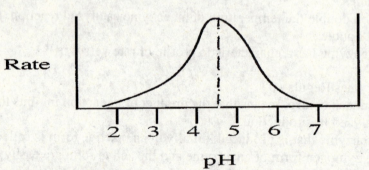

9. The (free) enzyme structure (without substrate bound to it) is different than when the substrate is bound to the active site.

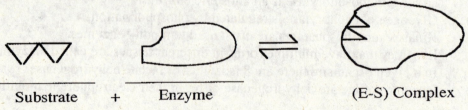

258

10. inorganic ions (such as Mg^{+2}, Zn^{+2}, etc). organic compounds, organic compounds

11. histidine, cysteine, aspartic acid, arginine, glutamic acid. acidic, basic, sulhydryl (thiol)

QUICK QUIZ

23.1

(a) False	(b) False	(c) True	(d) True
(e) True	(f) True	(g) True	

23.2

(a) True	(b) True	(c) False	(d) True
(e) False	(f) True		

23.4

(a) True	(b) False	(c) False	(d) True

23.6

(a) False	(b) True	(c) False	(d) True
(e) True	(f) True	(g) True	(h) True
(i) True	(j) False	(k) True	

One, if by land, or two, if by sea.
   - Paul Revere's Ride
     Henry Wadsworth Longfellow

24

Chemical Communications: Neurotransmitters and Hormones

CHAPTER OBJECTIVES

After you have studied the chapter and worked the assigned exercises in the text and the study guide, you should be able to do the following.

1. Tabulate the two general types of chemicals involved in intercellular communication, indicating (1) their general make up, (2) what each acts on, (3) the time frame of their action, (4) the cells or organs which are involved in the communication and (5) whether secondary messengers are involved in the process.

2. Explain the role of the axons, dendrites, the synapse, synaptic vesicles and receptor sites in the neurotransmission by acetylcholine.

3. Describe how acetylcholine bound to postsynaptic receptor sites is removed normally in the cell and how muscle relaxants or nerve gases can prevent this.

4. Discuss the action of drugs that function as agonists and antagonists on cellular receptors.

5. Discuss the role and mechanism of action of ligand-gated ion channels.

6. Outline the interrelationship between chemical messenger, membrane receptor, G-proteins and the enzyme, adenylate cyclase and the ultimate target.

7. Distinguish between the H_1 and H_2 histamine receptors in terms of their location, function and targeted treatment in disease.

8. Outline the mechanism of action and the final inactivation of adrenergic neurotransmitters by monoamine oxidases (MAO).

9. List the three types of hormones, their general make up and the types of action that each stimulates.

10. List a number of hormones, the gland that secretes the hormone and the specific action or function of the hormone (consult Table 24.2 in text).

11. Define the important terms and comparisons in this chapter and give specific examples where appropriate.

IMPORTANT TERMS AND COMPARISONS

Receptor Proteins
Neurons, Axons, Dendrites
Agonists and Antagonists
Endocrine Gland and Hormones
Adenylate Cyclase and cAMP
Neurotransmitter Classes
Cholinergic, Amino Acid, Adrenergic
    Peptidergic and Steroid
Acetylcholine and Acetylcholinesterase
Excitatory and Inhibitory Neurotransmitters
Phosphodiesterase
Calmodulin
L-Dopa and Dopamine
Alzheimer's Disease, β-amyloid Plaques & Tau
Histamine and Histamine Receptors, H_1 and H_2
Transporters and Reuptake Process
Enkephalins
Botox (Botulin Toxin)
Histidine and Histamine
Diabetes Mellitus and
    Non-Insulin-Dependent Diabetes
Steroid Hormones
Bisphenol A (BPA)
Insulin and Glucagon in Contol of Blood Glucose Levels

Chemical Messengers and
    Secondary Messengers
Neurotransmitters and Hormones
Synapse, Presynapse Site, Postsynapse
    Site and Vesicles
Hormone Classes
Adrenergic, Peptidergic and Steroid
Primary and Secondary Messengers
Transmembrane Proteins
Ligand-Gated Ion Channel Receptors
G-Protein-cAMP Cascade and
    Signal Transduction
Parkinson's Disease
Reversible and Irreversible Inhibitors of
    Acetylcholinesterase
Monoamine Oxidase (MAO)
Morphine
    Acetylcholine Transferase
NO and Viagra
Helicobacter pylori Bacteria and Ulcers
Ritalin and ADD (Attention Deficit
    Disorder)

SELF-TEST QUESTIONS

MULTIPLE CHOICE. In the following exercises, select the correct answer from the choices listed.

1. Decamethionium, which has a structure similar to that for acetylcholine, inhibits acetylcholinesterase. It is a

    a.   non-competitive inhibitor
    b.   chemical mediator
    c.   competitive inhibitor
    d.   toxic poison

2. Of the natural communication chemicals presented in the chapter, indicate the one which acts over the longest distance.

a. neurotransmitters   c. c-AMP

b. steroid hormones   d. NO

3. Acetylcholinesterase hydrolyzes which bond in acetylcholine, $H_3CCOCH_2CH_2N(CH_3)_3^+$?

$$\underset{O}{\overset{\|}{C}}$$

a. $H_3C\underset{\|}{\overset{\downarrow}{—}}C$ c. $H_3CCO\underset{\|}{\overset{\downarrow}{—}}C$

        O   O

b. $H_3C\text{-}C\underset{\|}{\overset{\downarrow}{—}}O$ d. $\text{-}H_2C\overset{\downarrow \;\;+}{—}N(CH_3)_3$

        O

4. H_1 receptors found in the respiratory tract are blocked by the interaction of

a. insulin   c. enkephalins

b. L-DOPA   d. antihistamines

5. The second messenger in the action of monoamine neurotransmitters is in most cases

a. d-AMP   c. AMP

b. c-AMP   d. ATP

6. The administration of L-Dopa can reverse the symptoms of Parkinson's disease. Dopamine is produced from L-Dopa by a

a. hydrolysis reaction   c. decarboxylation reaction

b. oxidative decarboxylation reaction   d. oxidative reaction

7. Protein kinases function as

a. phosphate transfer enzymes   c. secondary messengers

b. structural membrane proteins   d. ion translocating proteins

8. Neurotransmitters are stored in vesicles in the

a. axon   c. presynaptic site

b. dendrites   d. postsynaptic site

9. Glucagon is an example of a

a. peptide hormone   c. monoamine neurotransmitter

b. steroid hormone   d. receptor

10. An agonist is a molecule that

   a. stimulates antibody production
   b. competitively inhibits acetylcholinesterase
   c. activates receptor sites
   d. reduces pain

11. Steroid hormones produce their effect in the cell by

   a. binding to the membrane
   b. activating a protein kinase
   c. stimulating c-AMP production
   d. activating one or more genes

12. The hormone classes include all the molecules listed below, with the exception of

   a. peptides
   b. nucleotides
   c. steroids
   d. small molecules, such as amino acids

13. Indicate which of the following are general characteristics found in ligand-gated ion channels (receptors).

   a. They are multiple subunit proteins.
   b. They are transmembrane proteins.
   c. Ligand binding opens the interior channel to permit ion flow into and out of the cell.
   d. The flow of ions through the ion channel stimulates ligand binding.

COMPLETION. Write the word, phrase or number in the blank space in answering the question.

1. Complete the following reactions, indicating the enzyme that catalyzes the reaction and writing structural formulas for products and/or reactants.

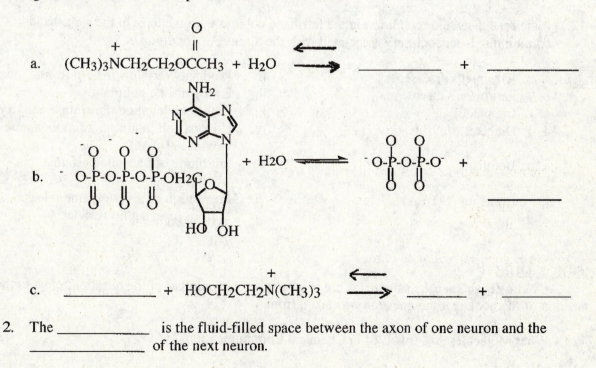

2. The _____ is the fluid-filled space between the axon of one neuron and the _____ of the next neuron.

3. A neurotransmitter and a receptor combine very specifically together. This interaction has been likened to the _____ _____ _____ model for enzyme-substrate interactions.

4. Food poisoning from botulism toxin deprives cholinergic nerves of their neurotransmitters. This indicates that the release of _____ from the presynaptic vesicles is blocked.

5. To degrade the neurotransmitter, acetylcholine, in the postsynaptic receptor site, the enzyme, _____ , acts on acetylcholine to _____ it to acetic acid and choline.

6. Monoamine neurotransmitters, such as norepinephrine, form a membrane bound _____ - _____ like complex, which activates the enzyme _____ that produces the secondary messenger _____ from ATP.

7. GABA is the abbreviation for _____ .

8. The _____ end of methionine enkephalin in similar to a segment of the alkaloid painkiller, _____ . Refer to Figure 24.5 in text.

9. Antidiabetic drugs are useful for diabetic patients who have insufficient numbers of _____ _____ on their target cells.

10. MAO is the abbreviation for _____ , which has the function of _____ monoamine neurotransmitters by _____ them to an _____ functional group.

11. The reaction type that converts histidine to histamine is the same type that converts L-Dopa to dopamine. This type of reaction is called a _____ reaction.

12. Match the disease or condition in the left-hand column with the term in the right-hand column that is most closely associated with the disease or condition.

a.	Alzeimer's disease	i.	Produces very low insulin levels
b.	Parkinson's disease	ii.	Estrogen-like pollutant
c.	Impotency	iii.	Low acetylcholine transferase activity
d.	Diabetes Mellitus, type I	iv.	L-Dopa and inhibitors of monoamine oxidase (MAO)
e.	Botulism	v.	Inhibitors of phosphodiesterase activity
f.	Biphenol A (BPA)	vi.	Prevention of acetylcholine release from presynaptic vesicles

QUICK QUIZ
Indicate whether the statements are true or false. The number of the question refers to the section of the book that the question was taken from.

24.1 What Molecules Are Involved in Chemical Communications?

264

(a) Receptors are peripheral protein molecules in loose contact with the cell membrane.

(b) Chemical messengers interact with receptors.

(c) Secondary messengers carry the original message inside the cell and amplify its effect.

(d) An agonist is a drug that blocks a receptor and interferes with its function.

24.2 How Are Chemical Messengers Classified as Neurotransmitters and Hormones?

(a) Neurotransmitters are released by a nerve cell and travel across a gap called a synapse to a receptor on a neighboring cell.

(b) Hormones differ from neurotransmitters in their physiological function, but not in their chemical nature.

(c) Chemical messengers affect the permeability of membranes but not the activity of enzymes.

(d) All neurotransmitters are amino acids or derivatives of amino acids.

24.3 How Does Acetylcholine Act as a Messenger?

(a) When acetylcholine binds to its receptor, a channel opens to allow passage of Ca^{2+} ions.

(b) In transmission of a nerve impulse, acetylcholine is released from vesicles into the synapse.

(c) When acetylcholine is released from the site for it on the receptor, it is hydrolyzed to acetate and choline.

(d) Inhibitors of the enzyme acetylcholinesterase, which hydrolyzes acetylcholine, do not have any marked physiological effect.

24.5 What Are Adrenergic Messengers?

(a) Adrenergic messengers tend to be monoamines.

(b) Secondary messengers tend not to play a role in the mode of action of adrenergic messengers.

(c) The term "signal transduction" refers to transmission of the message without amplification.

(d) Activation of G-protein eventually leads to the production of cyclic AMP.

(e) Cyclic AMP represses the production of secondary messengers.

(f) Adrenergic messengers are degraded by reactions catalyzed by the class of enzymes known as monoamine oxidases.

ANSWERS TO SELF-TEST QUESTIONS

MULTIPLE CHOICE

1.	c	7.	a
2.	b	8.	c
3.	b	9.	a
4.	d	10.	c
5.	b	11.	d
6.	c	12.	b
		13.	a, b, c

COMPLETION

1. a. acetylcholinesterase; H_3CCO_2H + choline
    b. adenylate cyclase; c-AMP (see figure in text, section 24.5)
    c. $H_3CC\overset{\overset{\displaystyle O}{\|}}{}\text{-SCoA}$; acetylcholine transferase; $H_3CCOCH_2CH_2\overset{+}{N}(CH_3)_3$ + CoA-SH, with the carbonyl $\overset{}{\underset{\|}{O}}$

2. synapse; dendrites (or cell body)
3. lock-and-key
4. acetylcholine
5. acetylcholinesterase; hydrolyze
6. hormone-receptor; adenylate cyclase; c-AMP
7. γ-aminobutyric acid
8. N-terminal end; morphine
9. insulin receptors
10. monoamine oxidase, inactivating, oxidizing, aldehyde
11. decarboxylation
12. a. iii
    b. iv
    c. v
    d. i
    e. vi
    f. ii

QUICK QUIZ

24.1
| (a) False | (b) True | (c) True | (d) False |

24.2
| (a) True | (b) False | (c) False | (d) False |

24.3
| (a) False | (b) True | (c) True | (d) False |

24.5
| (a) True | (b) False | (c) False | (d) True |
| (e) False | (f) True | | |

266

What distinguishes a butterfly from
a lion, a hen from a fly, or a worm
from a whale is........
    -F. Jacob
     Science, 196, 1161 (1977)

25

Nucleotides, Nucleic Acids and Heredity

CHAPTER OBJECTIVES

Each one of us is unique in many ways. These specific characteristics result because each individual has a different genetic makeup (your genome). So although the large majority of the nucleotides in our DNAs are the same, a small fraction of the nucleotides are different. These differences become manifest as the genes in the DNA are transcribed into RNA and then translated into the many diverse enzymes and proteins (the real "workers" in the cell), that reveal important aspects of our individuality. In this chapter, we learn about the two types of nucleic acids - DNA and RNA. After first dissecting nucleic acids into their constituents, the general structure of these polynucleotides and the packaging of the DNA in nucleosomes and the higher order structures of chromosomes is examined. The process of replication by which DNA is duplicated with high fidelity is presented, together with critical mechanisms involved with repairing damaged DNA. The process of cloning a segment of DNA is presented, together with a technique called polymerization chain reaction (PCR), which has revolutionized and streamlined many aspects of cloning. These concepts will provide the basis for examining how selected genes are expressed in cells (Chapter 26) and lead to an understanding of the Central Dogma of Molecular Genetics. The "flow" of genetic information is briefly summarized on the following page and should be consulted during your study of Chapters 25 and 26.

Genetic Material		Genetic Message		Genetic Protein Product

DNA **Transcription** ⟶ **Translation**

mRNA ┄┄┄┄┄┄┄⟶ Protein

tRNA **Protein Synthesis**

RNA Synthesis rRNA

Replication
the

(DNA Synthesis)

a. Amino acid specific tRNAs bring in "activated form" of the amino acid. Correct base pairing between the anti-codon triplet (tRNA) and the codon (mRNA) occurs. The amino acid which becomes incorporated into the protein is defined by the genetic code.

b. rRNA combines with proteins to form ribosomes, on which translation occurs

Nucleus   **Cytoplasm or Cytosol**

The complexity of DNA is further revealed in the exon and intron segments of the genes in higher organisms and the repeating nucleosome subunits in the nuclear chromosome. These properties were first realized in the mid 1970's.

After you have studied the chapter and worked the assigned exercises in the text and the study guide, you should be able to do the following.

1. Explain the importance of hereditary information and its storage and expression in DNA.

2. Name and draw the structures of the organic bases (purines and pyrimidines), the nucleosides and the nucleotides that occur in DNA and RNA.

3. Draw a dinucleotide unit of DNA containing one purine and one pyrimidine base; point out the 5'- and 3'-ends, the glycosidic bond between the sugar and base and the phosphodiester backbone in the dinucleotide.

4. Draw the complementary base pairs, G≡C, and A=T, showing all atoms and the hydrogen bonding interactions.

5. In double-stranded DNA, indicate the relative positions (inside or outside on the double helix) of the aromatic bases, the sugar and phosphate units in this structure.

6. List the compositional, structural and functional differences between DNA and RNA.

7. Outline the main features of DNA replication and the replisome, indicating the role or significance of unwinding proteins, DNA polymerase, complementary base pairing, semi-conservative replication, 5'---->3' direction, Okazaki fragments and DNA ligase.

8. Describe the structure of a tRNA molecule, the position of the anticodon loop, the role of the 3'-end and the function of this RNA in the translation process.

9. Discuss the role of telomerase in DNA replication and its proposed role in immortality.

10. Define the terms gene, exons and introns as they relate to higher organisms.

11. Describe the composition and structure of the nucleosome and how it folds to results in higher order structure for the chromosome.

12. Explain the immediate goal of the Human Genome Project and its importance in the understanding of human biology.

13. Indicate the important new use of DNA fingerprinting in forensic science.

14. Discuss how pharmacogenomics will help tailor the prescription of drugs.

15. Discuss how DNA repair processes operate to eliminate DNA damage.

16. Define how apoptosis differs from necrosis and indicate the role of capsases and endonucleases in the apoptotic process.

17. Outline how the PCR technique can amplify DNA and indicate the role of temperature, DNA polymerase and primers play in the procedure.

18. Define the important terms and comparisons in this chapter and give specific examples where appropriate.

IMPORTANT TERMS AND COMPARISONS

Chromosomes and Genes
Nucleic Acids
Ribonucleic Acid (RNA)
Nucleus (of the cell)
Pyrimidine Bases; C, T, U
D-Ribose and D-Deoxyribose
Bases, Nucleosides and Nucleotides
Ester and Anhydride Bonds
3'-and 5'-OH Ends of Nucleotides
Anticancer Drugs and Chemotherapy
Complementary Base Pairs
DNA Double Helix
Histone Proteins and Nucleosomes
Chromosomes and Chromatin
One Gene-One Enzyme Hypothesis
mRNA, rRNA, tRNA and Ribozymes
snRNAs and snRNPs
Exons, Introns, Splicing of mRNA and Ribozymes
Satellite DNA

Heredity
Deoxyribonucleic Acid (DNA)
Transcription and Translation
Cytoplasm
Purine Bases; A, G
β-N-Glycosidic Bond
C-3- and C-5-Carbon on Sugar Ring
Ribonucleotides and Deoxyribonucleotides
A, AMP, ADP and ATP
Primary and Secondary Structure of DNA
AT and GC Hydrogen Bonding
Major and Minor Grooves in DNA
Nuclesomes, Solenoids, Loops, Minibands,
    and Higher Order Chromatin Structure
(Inheritable) Gene
Clover Leaf Structure of tRNA
miRNAs and siRNA
Ribosomes
Semiconservative DNA Replication

Origin of Replication and Replication Fork   3'-and 5'-OH Ends of DNA
Leading and Lagging DNA Strands   Okazaki Fragments
Continuous and Discontinuous Strand Synthesis   Replisomes
Telomeres and Telomerase   Somatic Cells and Germ Cells
DNA Polymerase   Okazaki Fragments
Histone Acetylase and Chromatin   DNA Fingerprinting and Electrophoresis
Pharmacogenomics and EM, PM and UEM   Human Genome and 30,000 Genes
CYP2D6 Gene and Cytochrome P-450 Enzyme   DNA Repair, BER and NER
AP Sites in DNA   Apoptosis and Necrosis
Cloning   Polymerase Chain Reaction (PCR)
Hybridization of DNA Strands   Restriction Enzymes
Human Genome Project

Components of the Replisome
Topoisomerase or Gyrases
DNA Polymerase
DNA Ligase
Helicase
Primase

SELF-TEST QUESTIONS

MULTIPLE CHOICE. In the following exercises, select the correct answer from the choices listed. In some cases, two or more choices will be correct.

1. The backbone in a nucleic acid is called a

    a. β–glycosidic bond   c. phosphodiester backbone
    b. sugar backbone   d. peptide backbone

2. The coding sequences of a gene in DNA are called

    a. nucleosomes   c. exons
    b. introns   d. initiating factors

3. A linkage connecting the base and the deoxyribose is a

    a. peptide backbone   c. phosphodiester backbone
    b. β-N-glycosidic bond   d. other linkage

4. A nucleosome is made up of

    a. DNA and histone proteins   c. nucleosides
    b. rRNA and proteins   d. both DNA and RNA

5. Telomerase is an enzyme that

    a. is made of both an RNA and protein subunit
    b. replicates the telomeres at the ends of the chromosomes

    c.   replicates DNA in a semiconservative manner
    d.   can be used in PCR

6.   A feature which distinguishes ribose from deoxyribose is the absence of an OH at the

    a.   C-2-carbon   c.   C-1-carbon
    b.   C-3-carbon   d.   C-5-carbon

7.   The hydrogen bonding between a G and C in forming a GC base pair in DNA involves

    a.   atoms on or in the six-membered rings of both guanine and cytosine bases
    b.   atoms on or in the six-membered-ring of cytosine and the five-membered ring in
          guanine
    c.   only oxygen atoms in cytosine and nitrogen atoms in guanine
    d.   only hydrogen-bond donors in cytosine and only hydrogen-bond acceptors
          in guanine

8.   Adenine is the name of a

    a.   nucleoside   c.   nucleotide
    b.   purine base   d.   amino acid

9.   The sequence GCCCTGA has the A at the

    a.   5'-end   c.   can be either the 5'- or 3'- end
    b.   3'-end   d.   depends on size of DNA

10.   The 3'-end of a tRNA molecule

    a.   contains the anticodon loop
    b.   binds to a specific amino acid
    c.   has no known function
    d.   binds to rRNA

11.   The structure of cytosine is

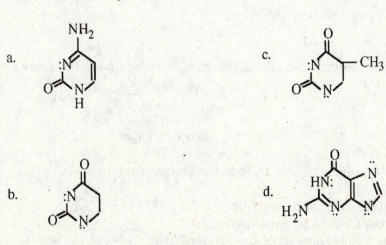

271

12. DNA fingerprinting requires the use of

    a. protein synthesis
    b. transcription factors
    c. restriction enzymes
    d. exons and introns

13. Okazaki fragments are

    a. formed in RNA transcription
    b. formed in DNA replication
    c. small RNA molecules
    d. small DNA fragments of about 1000 nucleotides each

14. A glycosidic bond in a nucleoside or nucleotide links the (deoxy)ribose sugar and which position on a purine base?

    a. N-1 atom   c. N-3 atom
    b. N-9 atom   d. C-8 atom

15. One of the complementary base pairs in DNA is

    a. AU   c. AT
    b. GC   d. GT

16. An RNA molecule has no

    a. uracil bases   c. cytosine bases
    b. thymine bases   d. guanine bases

17. The hydrogen bonding in an AT base pair involves atoms in the purine ring exclusively associated with the

    a. 6-membered ring
    b. 5-membered ring
    c. both the 5- and 6-membered ring
    d. none of the above

18. The process by which cells become specialized (neurons, muscle, etc.) in the development of a multicellular organism is called

    a. replication   c. cell division
    b. differentiation   d. genetic mutation

19. The single-stranded DNA primers used in PCR can be complementary to

    a. only the 3'-ends of the DNA to be amplified.
    b. only the 5'-end of the DNA to be amplified.
    c. either the 3'- or the 5'-ends of the DNA to be amplified.

d. both the 3'- or the 5'-ends of the DNA.

20 Caspases are

a. nucleases that cut at cytidine residues in DNA
b. enzymes used in cloning of DNA
c. enzymes that cleave proteins next to an aspartic acid residue
d. involved in programmed cell death called apoptosis

21. An RNA nucleoside does not contain a

a. heterocyclic base   c. ribose sugar
b. phosphate   d. amino acid

22. A nucleosome is composed of

a. DNA polymerase and other proteins and enzymes.
b. a sugar, heterocyclic base and phosphate.
c. 147 base pairs of DNA and eight histone protein.
d. RNA and protein.

23. It is thought that chromatin structure must be opened up before replication can occur.
     As part of this process,

a. histone deacetylase cleaves off an acetyl group from a serine residue in histone
    proteins.
b. histone acetylase puts an acetyl group onto a lysine reside in a histone protein.
c. a phosphate group is put on a histidine residue in a histone protein.
d. the histone proteins become glycosylated with a number of sugar groups.

24. The RNA that affects gene expression and has been used to "knock out" genes is:

a. siRNA   c. snRNA
b. mRNA   d. tRNA

COMPLETION. Write the word, phrase or number in the blank space in answering the question.

1. The DNA molecule in humans (the human genome) is thought to have at least
    _____ genes.

2. Replication of DNA in eukaryotic cells occurs in the _____ of the cell, while
    translation occurs in the _____.

3. There are _____ hydrogen bonds between adenine and its (structurally)
    complementary base, _____.

4. A nucleic acid that exhibits enzymatic activity is called a _____ .

5. An RNA-protein complex that exhibits enzymatic activity is called _____ and is associated with the addition of about 50-100 nucleotides from the _____ of chromosomes at each cell division.

6. Amino acids are bonded to the _____ end of tRNA on the _____ unit.

7. Replication of the new DNA strand occurs in a _____ direction, while the DNA polymerase is traveling in the _____ direction on the DNA template strand.

8. _____ are composed of rRNA molecules and proteins and are the units on which protein synthesis occurs in the _____ of the cell.

9. The organic bases in DNA reside on the interior of the double helix and stabilize the structure by _____ interactions.

10. _____ bases are made up of two fused rings, while _____ bases have only a six-membered heterocyclic ring.

11. The sequence of one strand of double stranded DNA is 5'-AGGAGCTTCG-3'. The sequence of the complementary strand is _____.

12. In BER (repair process), which is an acronym for _____ _____ _____, the enzyme that recognizes DNA damage and cleaves off a base is called _____ _____.

13. Match the scientist(s) in the left-hand column with the accomplishment on the right with which he (they) is most closely associated.

a.	Okazaki	(i)	Proposed the structure of DNA
b.	J. Watson & F. Crick	(ii)	Proposed DNA contains the genetic information
c.	K. Mullis		
d.	O. Avery	(iii)	X-ray studies of DNA
e.	R. Franklin & M. Wilkins	(iv)	Proposed one gene-one enzyme hypothesis
f.	G. Beadle & E. Tatum		
g.	E. Chargaff	(v)	Discovered that in DNA, # of A = # of T # of G = # of C
		(vi)	Discovered PCR technique
		(vii)	Discovered small segments of DNA made in DNA replication

QUICK QUIZ

Indicate whether the statements are true or false. The number of the question refers to the section of the book that the question was taken from.

25.2 What Are Nucleic Acids Made Of?
(a) Nucleic acids are so named because purines and pyrimidines are acidic.
(b) The monomer unit of nucleic acids is the purine.
(c) The bases that are commonly part of nucleic acids are adenine, guanine, cytosine, thymine, and uracil.

(d)   Thymine is found in DNA, while uracil is found in RNA.

(e)   A nucleoside is composed of a base, ribose or deoxyribose and a phosphate.

(f)   The only difference between ribose and 2-deoxyribose is the presence or absence of a hydroxyl group at position 2.

(g)   A nucleotide can have one, two or three phosphate groups.

25.3 What Is the Structure of DNA and RNA?

(a)   Nucleic acids are polymers of nucleotides.

(b)   Sequences of nucleotides are listed from 5' end to 3' end.

(c)   Chargaff's rule states that in DNA the amount of A + T always equals the amount of G+C.

(d)   A and T are called complementary base pairs.

(e)   The highest level of structure in eukaryotic DNA is called a histone.

25.4 What Are the Different Classes of RNA?

(a)   Ribosomal RNA is the largest of the RNA types.

(b)   Messenger RNA is the longest lived of the RNA types.

(c)   In the process of translation, tRNA is the interpreter between the mRNA and the protein.

(d)   Ribozymes are RNAs that exhibit catalytic activity.

(e)   All six types of RNA are required to make a protein.

25.6 How Is DNA Replicated?

(a)   DNA is synthesized continuously on both strands during replication.

(b)   DNA is synthesized in the 3' to 5' direction.

(c)   The basic chemical reaction of DNA synthesis is a nucleophilic attack.

(d)   Acetylation and deacetylation are important in control of replication.

(e)   Synthesis on the lagging strand produces Okazaki fragments.

(f)   DNA replication is semiconservative.

25.8 How Do We Amplify DNA?

(a)   Polymerase Chain Reaction is an automated process for duplicating DNA.

(b)   PCR can be accomplished when we know nothing about the sequence of DNA.

(c)   PCR requires a heat-stabile form of DNA polymerase.

(d)   One primer sequence is needed to run PCR.

(e)   PCR involves cycles of heating and cooling the DNA.

ANSWERS TO SELF-TEST QUESTIONS

MULTIPLE CHOICE

1.	c	12.	c
2.	c	13.	b
3.	b	14.	b
4.	a	15.	b, c
5.	a, b	16.	b
6.	a	17.	a
7.	a	18.	b
8.	b	19.	a

275

9.	b		20.	c, d
10.	b		21.	b
11.	a		22.	c
			23.	b
			24.	a

COMPLETION

1. 30,000
2. nucleus, cytoplasm (or cytosol)
3. two, thymine
4. ribozyme (or catalytic RNA)
5. telomerase, teleomers
6. 3', adenosine
7. 5'-->3', 3'-->5'
8. Ribosomes, cytoplasm
9. hydrophobic
10. Purine, pyrimidine
11. 3'-TCCTCGAAGC-5'
12. base excision repair, DNA glycosylase
13. a. (vii)
    b. (i)
    c. (vi)
    d. (ii)
    e. (iii)
    f. (iv)
    g. (v)

QUICK QUIZ

25.2

(a) False	(b) False	(c) True	(d) True
(e) False	(f) True	(g) True	

25.3

(a) True	(b) True	(c) False	(d) True
(e) False			

25.4

(a) True	(b) False	(c) True	(d) True
(e) False			

25.6

(a) False	(b) False	(c) True	(d) True
(e) True	(f) True		

25.8

(a) True	(b) False	(c) True	(d) False
(e) True			

It is the amazement of selective gene expression
that destines certain cells to become liver cells,
others become heart cells, etc.- all remarkably orchestrated
to make you, me and our neighbor.
                -Just thoughts

26

Gene Expression and Protein Synthesis

Each one of us is unique. However, the liver in every one of us functions in the same way, likewise the heart, the colon, etc. These characteristics result from the finding that although every cell contains the same genetic material, DNA, with its many genes, only a small select collection of these genes is expressed into proteins in any particular cell type (tissue). This chapter permits us to begin to understand these remarkable happenings within us.

CHAPTER OBJECTIVES

After you have studied the chapter and worked the assigned exercises in the text and the study guide, you should be able to do the following.

1. Consider a gene on DNA and go through the details of the step-wise processes of transcription and translation (gene --> protein).

2. Outline the transcription process and the role of transcription factors, mRNA and the codon units in the transfer of genetic information.

3. Describe the structure of a tRNA molecule, the position of the anticodon loop, the role of the 3'-end and the function of this RNA in the translation process.

4. Describe the location and role of rRNA in the translation process.

5. Define the genetic code and distinguish between the initiation codon, stop codons and degenerate codons.

6. Describe the processes of amino acid activation and then the initiation, elongation and termination steps in protein synthesis.

7. Describe the role of the A and P sites and the 40S and 60S ribosomes in translation.

8. Define the terms, gene, exons and introns as they relate to higher organisms.

9. Describe the post-transcriptional processing that occurs in the nucleus to convert the transcript to a mature mRNA that is transported to the cytoplasm for protein synthesis.

10. Explain how a mutation can be produced during DNA replication or by chemicals called mutagens.

11. Explain the use of plasmids, restriction endonucleases and Escherichia coli in recombinant DNA technology.

12. Describe the role of the promoter, regulatory sequences and regulator transcription factors in the regulation of the expression of eukaryotic structural genes.

13. Indicate the function of chaperone proteins and the proteasome in post-translational controls.

14. Explain the difference between a proto-oncogene and an oncogene and indicate its importance in the life of the cell.

15. Define the important terms and comparisons in this chapter and give specific examples where appropriate.

IMPORTANT TERMS AND COMPARISONS

Central Dogma of Molecular Biology
Gene Expression, Transcription and Translation
Template Strand, (-) Strand and Antisense Strand
Complementary Base Pairing-AT and GC
Ribonucleic Acid (RNA)
One Gene-One Enzyme Hypothesis
Initiation, Elongation and Termination
Gene Promoter
Transcription Factors
Exons and Introns
Yeast RNA Polymerase II
Svedberg Unit, S
Codon (Triplet Sequence) in mRNA
Genetic Code
Initiation and Termination Codons
Clover Leaf Structure for tRNA
Aminoacyl-tRNA Synthetase (AARS)
40 S and 60 S Ribosomes
Translation: Activation, Initiation, Elongation and Termination

Nucleus (of the Cell)
RNA Polymerase I, II and III
Coding Strand, (+) Strand and Sense Strand
Helicase
mRNA, rRNA and tRNA
Initiation and Termination Signals
Structural and Regulatory Genes
Consensus Sequences and Initiation Signals
Termination Sequence
mRNA Transcript and Mature mRNA
Post-transcriptional Processes
(Almost) Universal Genetic Code
Anticodon Recognition Site in tRNA
The Second Genetic Code
Degenerate or Multiple Codes
Activated Amino Acid
Amino Acid-tRNA
$tRNA^{fmet}$
P and A (Acceptor) Sites on the Ribosomes
Peptidyl Transferase

278

Selenocysteine and tRNA^{sec} Stop Codons

Translocation and Termination   Shine-Dalgarno Sequence

DNA Virus and RNA Virus   Antibiotics and Antiviral Agents

Retrovirus (RNA) and the AIDS Virus   Protease Inhibitors for AIDS Treatment

Post-Transcriptional Processing Events   Gene Regulation

     mRNA 5'-cap (7-mG)   Promoter, Enhancer and Silencer Sequences

     mRNA 3'-cap (Poly A Tail)   Genes

     Cutting out Introns & Splicing Exons   Transcription Regulatory Factors

Transcriptional Activation: Initiation, Elongation   Regulatory Sequences and Structural Genes

     and Termination   General Transcription Factors (GTFs)

TATA Box in Promoters   Preinitiation Complex

Protein Kinases   Enhancers and Response Elements

CREB: cAMP Response Element Binding Protein   CBP: CREB Binding Protein

Metal-Binding Fingers, Helix-Turn-Helix   Mutagens and Mutations

     And Leucine Zipper Transcription Factors   Mutagens and Carcinogens

Chaperones   Proteasomes

Proto-oncogenes and Oncogenes   Tumor Suppressor Gene and p53

Recombinant DNA   Escherichia coli (E. coli) Bacterium

Plasmids   Restriction Endonuclease

"Sticky Ends"   DNA Ligase

Genetic Engineering   Gene Therapy

Adenosine Deaminase Gene (ADA)   Severe Combined Immune Deficiency

Delivery Methods for Human Gene Therapy   (SCID)

     Ex Vivo   Viral Delivery Vector

     In Vivo   Maloney Murine Leukemia Virus (MMLV)

        Adenovirus

SELF-TEST QUESTIONS

<u>MULTIPLE CHOICE</u>. In the following exercises, select the correct answer from the choices listed. In some cases, two or more choices will be correct.

1. The central dogma of molecular biology states that

    a.   DNA is a double stranded helix
    b.   Information flows from DNA to RNA to proteins
    c.   Gene expression is different in all tissues
    d.   DNA replication is the key event in cell division

2. The coding sequences of a gene in DNA are called

    a.   nucleosomes   c.   exons
    b.   introns   d.   initiating factors

3. An RNA virus contains which of the following

    a.   DNA   c.   proteins
    b.   RNA   d.   carbohydrates

4. The RNA molecules that are translated into proteins are

   a. tRNA   c. mRNA
   b. rRNA   d. ribozymes

5. The three-letter nucleotide sequence that is listed for the genetic code is the sequence that is found on

   a. codon sequence of mRNA   c. anti-codon sequence of tRNA
   b. sequence on DNA   d. codon sequence on rRNA

6. In protein synthesis, the initial amino acid is always

   a. glycine   c. methionine
   b. cysteine   d. alanine

7. Translation occurs

   a. on the ribosomes in the cytoplasm.
   b. with the first amino acid being the N-terminal residue.
   c. on the ribosomes in the nucleus.
   d, with the first amino acid being the C-terminal residue.

8. In the activation of the amino acid prior to translation, how many high energy phosphate bonds are cleaved ?

   a. none   c. two
   b. one   d. three

9. The 3'-end of a tRNA molecule

   a. contains the anticodon loop
   b. binds to a specific amino acid
   c. has no known function
   d. binds to rRNA

10. The bond that links the amino acid residue to AMP is an

   a. ether bond   c. ester bond
   b. acid anhydride bond   d. amide bond

11. The initiating codon in protein synthesis is

   a. GGG   c. AUG
   b. GUA   d. ATG

12. Antibiotics are used to

    a. kill viruses   c. bind to mutagens
    b. inhibit DNA replication   d. kill bacteria

13. The ribosome subunit(s) that contains the P and A sites is (are)

    a. 60S   c. both the 40S and 60S
    b. 40S   d. found on the tRNA

14. The codons that code for the incorporation of the amino acid, cysteine, into a protein is
    (are)

    a. UUU   c. UGU
    b. UAG   d. UGC

15. The interactions in transcription often result in DNA forming a loop of DNA. In these
    cases, indicate the types of proteins (far from the promoter) that often help in the DNA
    looping by interacting with protein factors in the promoter.

    a. enhancer binding proteins   c. termination factors
    b. RNA polymerase   d. peptidyl transferase

16. In protein synthesis, the mRNA initially binds to the

    a. 30 S ribosomal subunit   c. 100 S nucleosome
    b. 40 S ribosomal subunit   d.   60 S ribosome subunit

17. A mRNA molecule contains no

    a. uracil bases   c. cytosine bases
    b. thymine bases   d. guanine bases

18. General transcription factors bind to the

    a. promoter site   c. A site
    b. structural gene   d. ribosome binding site

19. An amino acid is activated by reaction with

    a. tRNA   c. ATP
    b. synthetase   d. methionine

20. Restriction endonucleases have the property that they

    a. can ligate DNA molecules
    b. are required for termination of transcription
    c. cleave DNA in a sequence-specific manner

d. bind to regulatory sites in promoters

21. Proteins that contain metal-binding sites often bind to:

a. Zn^{+2}

c. H^{+}

b. Na^{+}

d. Fe^{+2}

22. HIV patients are often treated with a three-drug cocktail that contains

a. protease activator

c. nucleoside analogs

b. protease inhibitors

d. transcriptional inhibitors

23. A tumor suppressor gene that is mutated in about 40% of all human cancers is

a. EGF

c. p100

b. p53

d. RNA polymerase

24. Misfolded proteins are digested and eliminated by:

a. restriction endonucleases

c. proteasome

b. peptidyl transfererase

d. chaperone proteins

25. A number of transcription factors have been discovered that are identified by the presence of the following features in these proteins:

a. leucine zipper

c. metal-binding-fingers

b. helix-turn-helix motifs

d. all of these features

COMPLETION. Write the word, phrase or number in the blank space in answering the question.

1. The _____ site on the ribosome is the site at which the incoming tRNA binds, while the _____ site is the site where the tRNA binds that contains the growing peptide chain.

2. A mutation by a chemical or by radiation occurs directly on _____, but the result is then transmitted to both the intermediate _____ and the final _____ product.

3. Plasmids are small, _____ DNAs found in _____ cells. They are useful in the _____ of genes.

4. Human insulin is one of the first proteins to be produced by ___ ____ _____ (three words).

5. It has been found that all _____ are mutatgenic, but not all mutagens are _____.

6. All amino acids bind to the _____ end of tRNA on the _____ unit.

7. Referring to Table 26.1 for the Genetic Code in the text, write the amino acid sequence for the protein produced from the mRNA shown below: Indicate the N- and C-terminal ends of the protein.

AUGUUGCACAACGGGGCGGUGGUGUAA

8. After combining the DNA of a plasmid that has sticky ends with another fragment of DNA with the same sticky ends, the enzyme _____ _____ can covalently splice the two DNA fragments together.

9. Ionizing radiation, gamma rays and chemicals, including benzene and vinyl chloride, are known to be _____.

10. The sequence of one strand of double stranded DNA is 5'-AGGAGCTTCG-3'. The sequence of the RNA strand produced from transcription on this strand is _____.

11. Synthesis of RNA (transcription) occurs in a _____ direction as RNA. The polymerase moves along the complementary DNA strand in a _____ direction.

12. The development of cancer may occur when a normal gene, called a _____, is altered or _____ . The resulting mutated gene is referred to as an _____.

13. Match the scientist(s) or term in the left-hand column with the most closely associated term in the right-hand column.

a. AIDS	(i)	UAA
b. A stop codon	(ii)	Determined the genetic code
c. Needed in protein synthesis initiation	(iii)	Caused by a retrovirus infection
d. p53	(iv)	Cuts DNA at a very specific sequence
e. Restriction endonuclease	(v)	Proposed the one gene-one enzyme hypothesis
f. G. Beadle & E. Tatum	(vi)	Most commonly mutated gene in human cancers
g. M. Nirenberg	(vii)	Shine-Dalgarno sequence

QUICK QUIZ
    Indicate whether the statements are true or false. The number of the question refers to the section of the book that the question was taken from.

26.1 How Does DNA Lead to RNA and Protein?
    (a)   Gene regulation comprises transcription only.
    (b)   DNA always leads to RNA, which always leads to protein.
    (c)   Transcription leads to protein when the RNA produced is mRNA.
    (d)   Translation is the process whereby mRNA is used to direct the synthesis of protein.
    (e)   Reverse transcriptase is an enzyme found in retroviruses.

26.2 How Is DNA Transcribed into RNA?
    (a)   When a gene is transcribed, both strands of DNA are used to make RNA.
    (b)   The DNA strand that is used by RNA polymerase is called the template strand.
    (c)   All regulatory sequences described on DNA are given as coding strand sequences.
    (d)   RNA polymerase transcribes both strands of DNA, one after another.

(e)   RNA polymerase transcribes only the structural part of the gene and not the regulatory region of the gene.

26.4 What Is the Genetic Code?
(a)   All proteins begin with methionine.
(b)   Methionine is initially the first amino acid of a peptide being synthesized.
(c)   There is only one codon for methionine.
(d)   All possible codons encode amino acids.
(e)   All amino acids have the same number of codons.
(f)   There can be as few as one or as many as 6 codons for a given amino acid.

26.5 How Is Protein Synthesized?
(a)   The four stages of translation are activation, initiation, elongation, and termination.
(b)   Activation requires an input of energy.
(c)   There are 64 types of tRNAs.
(d)   The amino acid is attached to the 5' end of tRNA during activation.
(e)   There are 3 sites on an intact prokaryotic ribosome.
(f)   The reaction that makes a peptide bond is a nucleophilic attack.
(g)   Peptide bond synthesis is catalyzed by a serine residue that makes a nucleophilic attack more favorable.
(h)   Incoming aminoacyl-tRNAs bind to the A site of the ribosome
(i)   In the translocation step, the ribosome moves on the mRNA by one nucleotide

26.9 What Is Gene Therapy?
(a)   Adenosine Deaminase deficiency can lead to Severe Combined Immunodeficiency.
(b)   The most common vector used for "*in vivo*" gene therapy is Maloney Murine Leukemia Virus.
(c)   With "*in vivo*" gene therapy, a patient is directly infected with a virus carrying the therapeutic gene.
(d)   Gene therapy is not approved for alteration of germ line cells (gametes).
(e)   Possible risks of gene therapy include strong reactions to the virus used to deliver the gene.

ANSWERS TO SELF-TEST QUESTIONS

MULTIPLE CHOICE

1.	b		13.	c
2.	c		14.	c, d
3.	b, c		15.	a
4.	c		16.	b
5.	a		17.	b
6.	c		18.	a
7.	a, b		19.	c
8.	c		20.	c

9.	b		21.	a
10.	b		22.	b, c
11.	c		23.	b
12.	d		24.	c
			25.	d

COMPLETION

1. A, P
2. DNA, RNA, protein
3. circular, bacterial (E. coli), cloning
4. recombinant DNA technology
5. carcinogens, carcinogenic
6. 3', adenosine
7. (N-terminal) Met-Leu-His-Asn-Gly-Ala-Val-Val (C-terminal)
8. DNA ligase
9. mutagenic and carcinogenic
10. 3'-UCCUCGAAGC-5'
11. 5'——>3', 3'——>5'
12. proto-oncogene, mutated, oncogene
13.
   a. iii
   b. i
   c. vii
   d. vi
   e. iv
   f. v
   g. ii

QUICK QUIZ

26.1

(a) False	(b) False	(c) True	(d) True
(e) True			

26.2

(a) False	(b) True	(c) True	(d) False
(e) True			

26.4

(a) False	(b) True	(c) True	(d) False
(e) False	(f) True		

26.5

(a) True	(b) True	(c) False	(d) False
(e) True	(f) True	(g) False	(h) True
(i) False			

26.9

(a) True	(b) False	(c) False	(d) True
(e) True			

"Whatever a cell does has to be paid
for in the currency of energy."
    - Szent-Gyorgyi

27

Bioenergetics:
How the Body Converts Food to Energy

CHAPTER OBJECTIVES

After you have studied the chapter and worked the assigned exercises in the text and the study guide, you should be able to do the following.

1. List the major organelles in a typical animal cell and indicate at least one important biological process that is carried out in each organelle.

2. List the (molecular) components contained in ATP, the coenzymes, NAD^+ and FAD, and acetyl CoA. Write out a general reaction in which each takes part and draw the part of each structure that is changed as a result of the reaction.

3. Distinguish generally between the catabolic and the anabolic process in terms of the consumption or production of ATP, NAD^+ and FAD.

4. Draw a clear schematic representation of a mitochondrion pointing out the outer membrane, inner membrane, intermembrane region, the matrix, cristae, and the location of the proton translocating ATPase.

5. State the major role of ATP in metabolic reactions.

6. Explain the role of NAD^+ & FAD in the citric acid cycle & in the electron transport chain.

7. Describe the role of acetyl CoA in the Krebs cycle.

8. Name the major species in the Krebs cycle.

9. Indicate the type of reaction involved at each step in the Krebs cycle and indicate (in words) an example of each.

10. Explain how the reduced coenzymes, NADH and $FADH_2$, are involved in oxidative phosphorylation and why each produces a different number of ATP molecules.

11. Determine the number of ATP molecules produced in the Krebs cycle and oxidative phosphorylation as a result of the oxidation of one acetyl group.

12. Describe the important features of the chemiosmotic hypothesis.

13. Describe the function of the proton translocating ATP-ase.

14. Explain the value of using an uncoupling agent in oxidative phosphorylation studies.

15. Give specific examples of the use of ATP to produce mechanical, electrical and chemical energy.

16. Describe the roles of the enzymes cytochrome P-450, superoxide dismutase and catalase in specific oxidative reactions.

17. Indicate the three major roles for oxygen in the cell.

18. Define the important terms and comparisons in this chapter and give specific examples where appropriate.

IMPORTANT TERMS AND COMPARISONS

Metabolism: Catabolism and Anabolism
Common Catabolic Pathway
Organelles
Riboflavin and Flavin
Translocator Outer Membrane Channels (TOM)
Coenzymes
Flavoproteins
Mitochondrion-Outer and Inner Membrane,
    Intermembrane Space, Matrix and Cristae
Reduced Coenzymes: NADH and FADH2
Activated C_2 Fragments or Acetyl Coenzyme A
Turn Over Rate for ATP
High-Energy Phosphate Anhydride Bond
Nicotinamide and Riboflavin
Mercaptoethylamine
Dehydration Reaction
Tertiary and Secondary Alcohol
GTP and ATP: Energetically Equivalent
Oxidative Decarboxylation Reaction
Coenzyme Q (CoQ)
Uncoupling Agents
Oxidative Phosphorylation
Proton Translocating ATP-ase
Cytochrome P-450
Superoxide (Anion), O_2^-

Metabolic or Biochemical Pathways
Energy Carrier Molecule, ATP
Nucleus, Lysosomes, Golgi Bodies,
    Mitochondria
Translocator Inner Membrane Complexes
    (TIM)
H^+ and Electron-Transporting Molecules
NADH and NAD^+ Couple
FADH2 and FAD Couple
Oxidized Coenzymes: NAD^+ and FAD
Coenzyme A (CoA), Acetyl Group
    (CH3CO-) and Acetyl Coenzyme A
AMP, ADP and ATP
Pantothenic Acid
Krebs Cycle, Tricarboxylic Acid Cycle and
    Citric Acid Cycle
Hans Krebs and Peter Mitchell
7.3 kcal/mol
Acetyl Group (CH3CO-)
Cytochrome c
Chemiosmotic Theory
Electron Transport Chain
Proton Gradient and Proton Channel
Protonphore
Superoxide Dismutase (SOD)

Catalase
Energy Yield

Glutathione
Energy Conversions

$$1/2 \; O_2 + 2 \; H^+ + 2 \; e^- = H_2O + Energy$$

FOCUSED REVIEW

I. Stages of Metabolism

Food Stuffs or Fuel Molecules

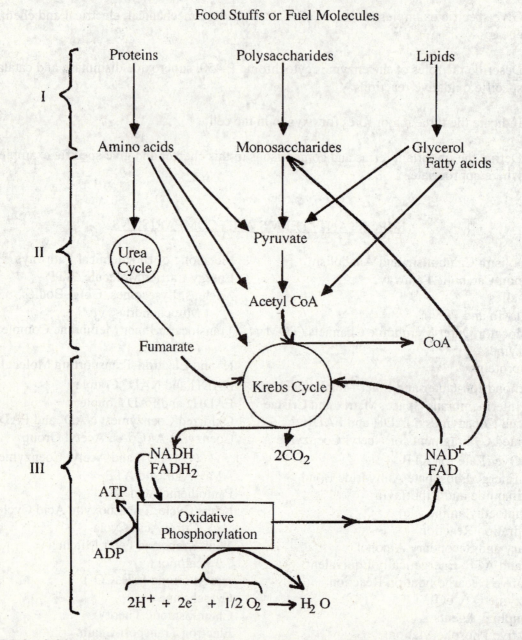

In the initial stage, the macromolecules in food are digested to monomers. Proteins are hydrolyzed to amino acids, polysaccharides to glucose and other monosaccharides and lipids to fatty acids and glycerol. In the second stage, the sugars, fatty acids, glycerol and many of the

288

amino acids are converted into the activated acetyl units (C_2 fragments) of acetyl coenzyme A. The nitrogen atoms in amino acids are processed in the urea cycle, while the deaminated carbon skeletons are processed directly or in a few steps into the Krebs cycle. The major focus in stage III involves the Krebs cycle and oxidative phosphorylation. In these final stages of the oxidation of "fuel" molecules, the C_2 fragment of acetyl CoA enters the Krebs cycle and is completely oxidized to $2\ CO_2$, reduced coenzymes NADH and $FADH_2$ are produced, along with one GTP molecule.

The reduced coenzymes, NADH and $FADH_2$, enter in the electron transport chain and are oxidized to NAD^+ and FAD, which are then available for recycling into the Krebs cycle and other catabolic pathways, including the glycolysis and the β-oxidation of fatty acids pathways (Chapter 27). As a result of the oxidation, hydrogen ions are expelled from the mitochondria and upon re-entry produce ATP from the phosphorylation of ADP. The electrons and the hydrogen ions now combine with O_2 to form water. Note that most of the ATP generated from the catabolic metabolism of the foodstuffs is formed in stage III.

II. Summary of Reactions

a. Hydrolysis (interconversion) of ATP to ADP + Pi

$$ATP\ =\ ADP\ +\ Pi$$

b. Half-reactions for coenzymes NAD^+ and FAD

$$NAD^+ + H^+ + 2\ e^- = NADH$$
$$FAD + 2\ H^+ + 2\ e^- = FADH_2$$

c. Overall reaction in Krebs cycle

$$H_3CCO\text{-}CoA\ +\ 2\ H_2O\ +\ 3\ NAD^+\ +\ FAD\ +\ GDP\ +\ Pi\ \longrightarrow$$

$$CoA\ +\ 2\ CO_2\ +\ \underbrace{3\ NADH\ +\ FADH_2}\ +\ GTP\ +\ 3\ H^+$$

| a waste | to electron | energy |
| product | transport chain | |

d. Overall reaction of oxidative phosphorylation

$$NADH\ +\ 3\ ADP\ +\ 1/2\ O_2\ +\ 3\ Pi\ +\ H^+\ =\ NAD^+\ +\ 3\ ATP\ +\ 4\ H_2O$$
$$FADH_2\ +\ 2\ ADP\ +\ 1/2\ O_2\ +\ 2\ Pi\ =\ FAD\ +\ 2\ ATP\ +\ 3\ H_2O$$

from
Krebs cycle

recycle   energy
back to
Krebs cycle

e. Oxidation of a C_2 (acetyl) fragment
$$C_2\ +\ 2\ O_2\ +\ 12\ ADP\ +\ 12\ Pi\ =\ 12\ ATP\ +\ 2\ CO_2$$

III. Citric Acid Cycle*

Step	Reaction	Type
1.	Acetyl CoA + oxaloacetate + H_2O $\longrightarrow$ citrate + CoA ($C_4H_2O_5$) ($C_6H_5O_7$)	Condensation reaction
2.	Citrate $\longrightarrow$ cis-aconitate + H_2O $\longrightarrow$ Isocitrate ($C_6H_5O_7$) ($C_6H_5O_7$) tertiary alcohol secondary alcohol	Isomerization, by way of a dehydration, followed by a hydration
3.	Isocitrate + NAD^+ $\longrightarrow$ α-ketoglutarate + NADH + CO_2 ($C_6H_5O_7$) ($C_5H_4O_5$) alcohol ketone	Oxidative decarboxylation
4/5.	α-ketoglutarate + NAD^+ + H_2O + GDP + Pi $\longrightarrow$ succinate + CO_2 + NADH + H^+ + GTP ($C_5H_4O_5$) ($C_4H_4O_4$) ketone acid	Oxidative decarboxylation
6.	Succinate + FAD $\longrightarrow$ Fumarate + $FADH_2$ ($C_4H_4O_4$) ($C_4H_2O_5$)	Oxidation (of succinate)
7.	Fumarate + H_2O $\longrightarrow$ malate ($C_4H_2O_4$) ($C_4H_4O_5$)	Hydration
8.	Malate + NAD^+ $\longrightarrow$ oxaloacetate + NADH + H^+ ($C_4H_4O_5$) ($C_4H_2O_5$) alcohol ketone	Oxidation (of malate)

290

*The Krebs cycle can be regarded as a catalytic cycle in which a 2-carbon fragment enters the cycle in the form of acetic acid (an acetyl group plus the OH from H_2O), is oxidized and leaves the cycle as CO_2.

The major species in each step of the cycle can be easily remembered by using the following saying as a mnemonic device. "**O**f **C**ourse, **I** **K**now **S**ome **F**amous **M**en". The first letter in each substrate - **O**xaloacetate, **C**itrate, **I**socitrate, α-**K**etoglutarate, **S**uccinate, **F**umarate and **M**alate begins with these same letters as in the saying.

Again, the overall reaction in the Krebs cycle is:

$$CH_3CO\text{-}CoA + 2\ H_2O + 3\ NAD^+ + FAD + GDP + Pi \quad \text{-----------}>$$

$$CoA + 2\ CO_2 + 3\ NADH + FADH_2 + 3\ H^+ + GTP$$

IV. Mitochondrion Components

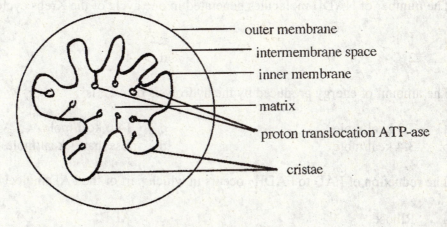

- outer membrane
- intermembrane space
- inner membrane
- matrix
- proton translocation ATP-ase
- cristae

SELF-TEST QUESTIONS

<u>MULTIPLE CHOICE</u>. In the following exercises, select the correct answer from the choices listed. In some cases, two or more answers will be correct.

1. $FADH_2$ transfers its electrons into the electron transport chain in the mitochondrion directly to

   a. ATP-ase
   b. Flavoprotein
   c. Q Enzyme
   d. Cytochrome c

2. Indicate which of the following processes does not consume ATP.

   a. Muscle contraction
   b. Flow of electrons into mitochondria through proton translocating ATP-ase
   c. Catabolism
   d. Pumping of K^+ into the cells by transport proteins

291

3. In the reaction in which succinate is converted to fumarate in the TCA cycle, succinate undergoes a

    a. Hydration   c. Isomerization
    b. Oxidation   d. Oxidative decarboxylation

4. Which of the following molecules are produced in the Krebs cycle?

    a. NAD^+ c. GDP
    b. CO_2 d. Oxaloacetate

5. The proton translocating ATPase is located where in the mitochondrion?

    a. Intermembrane space   c. Inner membrane/matrix
    b. Outer membrane   d. Outside the inner membrane

6. The number of NADH molecules generated in one cycle of the Krebs cycle is

    a. 1   c. 3
    b. 2   d. 4

7. The amount of energy produced by the hydrolysis of ATP is

    a. 7.3 kcal/mole   c. 14.6 kcal/mole
    b. 3.4 kcal/mole   d. less than 1 kcal/mole

8. The reduction of FAD to $FADH_2$ occurs in which part of the FAD molecule?

    a. Ribose   c. ADP
    b. Flavin   d. Nicotinamide

9. Indicate which of the following molecules are carriers of intermediates in metabolism.

    a. CoA   c. FAD
    b. ADP   d. Nicotinic acid

10. Of the following molecules, which contain high energy bonds?

    a. ATP   c. CO_2
    b. AMP   d. Isocitrate

11. The number of electrons, originating from the electron transport chain, that is needed to produce one mole of water is:

    a. 1   c. 3
    b. 2   d. 4

12. Niacin is another name for

    a.   Nicotinic acid   c.   Q enzyme
    b. α-ketoglutarate   d.   Nicotinamide

13. Which of the following terms is NOT another name for the Krebs cycle?

    a.   Tricarboxylic acid cycle
    b.   Electron transport chain
    c.   Oxidative decarboxylation
    d.   Citric acid cycle

14. The acetyl group in acetyl CoA is bonded to coenzyme A through what kind of a bond?

    a.   (C-C) bond   c.   (C-O) bond
    b.   (C-S) bond   d.   (C-N) bond

15. Which of the following molecules are NOT an electron-carrier in the mitochondrial membrane?

    a.   Cytochrome a_3 c.   Fumarase
    b.   Q enzyme   d.   Citrate synthetase

16. For each two C_2 fragment that enters the TCA (tricarboxylic acid) cycle, the total number of ATP and (GTP) molecules produced in the cell is

    a.   24   c.   10
    b.   22   d.   12

COMPLETION. Write the word, phrase or number in the blank or draw the appropriate structure
in answering the question.

1. A _____ is a compound that permits H^+ ions to pass through the _____ membrane passively.

2. The correct order for the enzymes, Q enzymes, FeS protein, cytochromes and flavoprotein in the electron transport chain is

    _____ , _____ , _____ , _____

3. Anabolism is associated with the _____ of molecules, while catabolism deals with the _____ of molecules.

4. _____ is the molecular unit which is common in FAD, NAD^+, ATP and coenzyme A.

5. For each NADH molecule, a total of _____ pairs of protons are pumped out of the mitochondrion into the _____ _____ , with the resultant production of _____ molecules of ATP.

6. Identify the 6 most important features of the mitochondrion

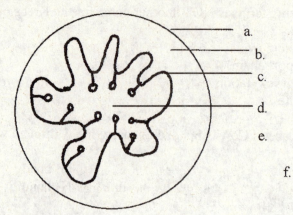

a. _____

b. _____

c. _____

d. _____

e. _____

f. _____

7. Considering the four intermediate carriers, _____ is a molecular unit unique to FAD, while _____ is a unique unit occurring in coenzyme A.

8. In the citric acid cycle, the tertiary alcohol in _____ must be converted to a secondary alcohol in _____ before the subsequent oxidation to produce a _____ (functional) group can occur.

9. The _____ hypothesis, proposed by Peter Mitchell states (summarize this in a short paragraph).

10. The reactions of the Krebs cycle occur in the _____ (an organelle).

11. The products of the reaction of acetyl CoA and oxaloacetate are _____ and _____ .

12. The reactions in the electron transport chain eventually lead to the reaction of O_2 with _____ and _____ .

13. NAD^+ is a coenzyme that can act as an _____ agent in the Krebs cycle. In the process of being converted to NADH, it becomes _____ and picks up _____ (a number) proton and _____ (a number) electrons.

14. The major organelles and a primary function of each is:
The _____ is where DNA replication takes place.
The removal of damaged cellular components occurs in the _____.
_____ _____ are responsible for the packaging and processing of proteins which are to be secreted from the cell.

15. The two critical intracellular enzymes that are responsible for decomposing the highly reactive superoxide anion are _____ _____ and _____ .

294

Indicate whether the statements are true or false. The number of the question refers to the section of the book that the question was taken from.

27.3 What Are the Principal Compounds of the Common Metabolic Pathway?
(a) AMP is the usual agent for the transfer of phosphate groups because its energy of hydrolysis is higher than that of other possible compounds.
(b) ATP contains a phosphoric-ester bond, in addition to phosphoric anhydride bonds as well.
(c) The electron-transfer agents in biological oxidation-reduction reactions contain derivatives of vitamins in the part of their structure that transfers electrons.
(d) Coenzyme A is an acetyl-transporting agent, but it differs from the electron-transport agent in not having an ADP core as part of its structure.

27.4 What Role Does the Citric Acid Cycle Play in Metabolism?
(a) Two steps of the citric acid cycle are oxidative decarboxylations.
(b) One step of the citric acid cycle produces ATP.
(c) The carbon dioxide we exhale comes from the two carbons of the acetyl group of acetyl-CoA in each round of the citric acid cycle.
(d) NAD^+ and FAD play roles as electron-transfer agents in the citric acid cycle.
(e) Feedback control regulates the citric acid cycle.
(f) Most of the energy yield of the citric acid cycle comes from indirect ATP production.

27.5 How Do Electron and H^+ Transport Take Place?
(a) The transfer of electrons and H^+ from the reduced coenzymes NADH and $FADH_2$ to oxygen requires the presence of large protein complexes in the inner mitochondrial membrane.
(b) The transfer of electrons and H^+ from the reduced coenzymes NADH and $FADH_2$ to oxygen requires the presence of mobile electron carriers.
(c) Two electrons are transported from one carrier to another at every stage of transport.
(d) Four protein complexes move H^+ across the inner mitochondrial membrane from the intermembrane space to the matrix.

27.6 What Is the Role of the Chemiosmotic Pump in ATP Production?
(a) ATP generation depends on maintaining a difference in H^+ concentration between the mitochondrial matrix and the intermembrane space.
(b) The proton translocating enzyme generates ATP by acting as if it were a rotor engine.
(c) All subunits of the ATP-generating enzyme are located in the inner mitochondrial membrane.
(d) The enzyme that produces ATP is also called proton translocator ATPase because it catalyzes both the forward and the reverse reactions of ATP production

ANSWERS TO SELF-TEST QUESTIONS

MULTIPLE CHOICE

1.	c	9.	a, c
2.	b, c	10.	a

3.	b	11.	b
4.	b, d	12.	a
5.	c	13.	b, c
6.	c	14.	b
7.	a	15.	c, d
8.	b	16.	a

COMPLETION

1. protonphore, mitochondrial
2. flavoprotein, FeS protein, Q enzyme, cytochrome
3. synthesis, breaking down
4. ADP
5. three, intermembrane space, three
6. Refer to the figure in Section IV of Focused Review
   a. outer membrane
   b. intermembrane space
   c. inner membrane
   d. matrix
   e. proton translocating ATP-ase
   f. cristae
7. flavin, pantothenic acid
8. citric acid, isocitrate, ketone
9. chemiosmotic, the energy in the electron transfer chain creates a proton gradient. There is a higher concentration of H^+ in the intermembrane space than inside the mitochondrial matrix. The protons pumped out of the matrix provide the driving force for the phosphorylation of ADP by driving the protons back into the mitochondrial matrix through the proton translocating ATPase. This enzyme manufactures the ATP.
10. mitochondrion
11. citrate, CoA
12. H^+, electrons
13. oxidizing, reduced, one, two
14. nucleus, lysosomes, Golgi bodies
15. superoxide dismutase (SOD), catalase

QUICK QUIZ

27.3
(a) False	(b) True	(c) True	(d) False

27.4
(a) True	(b) False	(c) False	(d) True
(e) True	(f) False		

27.5
(a) True	(b) True	(c) False	(d) False

27.6
(a) True	(b) True	(c) False	(d) True

"You are what you eat."
   - Unknown

28

Specific Catabolic Pathways:
Carbohydrate, Lipid, and Protein
Metabolism

CHAPTER OBJECTIVES

After you have studied the chapter and worked the assigned exercises in the text and the study guide, you should be able to do the following.

1. Name the major pathways by which energy is extracted from monosaccharides and fatty acids.

2. Discuss the three stages of glycolysis with regard to the energetics (ATP consumption or generation), the number of carbons in the molecular units and the reaction types involved.

3. Distinguish between anaerobic and aerobic glycolysis.

4. Indicate under what conditions pyruvate is converted to (a) acetyl CoA, (b) lactic acid or (c) ethanol. Write out the reaction for each process.

5. Tabulate the energy yield produced by reactions involved in the complete metabolism of glucose. Indicate the influence of the glycerol-1-phosphate or the aspartate-malate shuttle in the total energy yield.

6. Indicate the point in the catabolic process at which (a) glycerol and (b) fatty acids converge to a common pathway with monosaccharides.

7. Indicate the types of reactions in the β-oxidation spiral of fatty acids.

8. Specify the cellular location in which (a) glycolysis, (b) fatty acid activation and oxidation, (c) citric acid cycle, (d) oxidative phosphorylation and the (e) urea acid cycle occur.

9. Indicate the role of the pentose phosphate pathway in general anabolic metabolism and

nucleic acid metabolism, in addition to its indirect role in maintaining red blood cells.

10. Outline the role of phosphorylation in the regulation of fatty acid metabolism.

11. Compare the energy yields for the complete oxidation of (a) glucose, (b) glycerol and (c) a fatty acid.

12. Name the ketone bodies and explain why they are found in high concentrations in the blood under certain conditions.

13. Discuss the three stages of nitrogen metabolism.

14. Indicate the relationship between heme, biliverdin, bilirubin, and urobilin.

15. Define the important terms and comparisons in this chapter and give specific examples where appropriate.

IMPORTANT TERMS AND COMPARISONS

Food

Fat Storage Depots

Common Catabolic Pathway
   Citric Acid Cycle & Oxidative Phosphorylation

Glucose to Pyruvate or Lactic Acid

Pentose Phosphate Pathway

Glycerol

Energy Yield for Glucose

β-Oxidation Spiral (of Fatty Acids)

Carnitine Acyltransferase

Stearic Acid

Ketoacidosis

Acyl-CoA and Acetyl-CoA

Transamination and Oxidative Deamination

α-Ketoglutarate

Carbamoyl Phosphate

Ubiquitin, Proteasome and Protein Degradation

Phenylketonuria (PKU)

Jaundice and Bilirubin Levels

Glycerol-1-Phosphate Transport
   (Skeletal Muscles and Nerves)

Carbohydrates, Lipids and Proteins

Amino Acid Pool

Glycolysis

Aerobic Versus Anaerobic Pathway

Kinases, Isomerases and Aldolase

Glucose 6-Phosphate Conversion to
   Ribose and NADPH

Glucogenolysis: Glycogen to Glucose

Mitochondria

Thiokinase

Ketone Bodies: Acetoacetate, Acetone and
   β-Hydroxybutyrate

Urea Cycle: Production of Urea Form
   ATP, H_2O, CO_2 and NH_4^+

Ornithine and Citrulline

Fumarate: Linkage Between Urea and TCA

Amino Acids: Glucogenic and Ketogenic

Heme Catabolism: Heme, Biliverdine,
   Bilirubin and Urobilin

Ferritin

Aspartate-Malate Shuttle (Heart and Liver)

SELF-TEST QUESTIONS

MULTIPLE CHOICE. In the following exercises, select the correct answer from the choices listed. In some cases, two or more answers will be correct.

1. The initial reaction in the β-oxidation spiral of fatty acids produces

   a. acyl CoA
   b. acetyl CoA
   c. carbamoyl phosphate
   d. dihydroxyacetone phosphate

2. The final product of aerobic glycolysis is

   a. pyruvate
   b. acetyl CoA
   c. lactate
   d. ethanol

3. The number of ATP molecules produced on the complete metabolism of glucose in skeletal muscle is

   a. 24
   b. 36
   c. 12
   d. 38

4. Anaerobic glycolysis occurs in the

   a. mitochondria
   b. cytosol
   c. cytosol and mitochondria
   d. extracellular fluid

5. Ketone bodies are produced in conditions of a

   a. high glucose supply
   b. low amino acid pool
   c. low glucose supply
   d. missing fructose-1-phosphate aldolase

6. The heme cofactor in hemoglobin contains

   a. Fe^{+2}
   b. globin protein
   c. carbohydrate
   d. glycerol

7. The first reaction involved in nitrogen catabolism of amino acids is a(n)

   a. oxidative deamination
   b. reduction
   c. hydration
   d. transamination

8. The final step in the urea cycle involves the

   a. splitting of argininosuccinate
   b. oxidation of glutamate
   c. hydrolysis of arginine
   d. transamination of glutamate

9. The order of the caloric value for carbohydrates, proteins and fats is

   a. protein = fat = carbohydrate
   b. carbohydrate > fat > protein
   c. fat > carbohydrate > protein
   d. protein > fat > carbohydrate

10. Pyruvate undergoes an oxidative decarboxylation with CoA to produce acetyl CoA and

    a.   lactate   c. CO_2

    b.   glutamate   d. NAD^+

11. Which of the following molecules (in the form of reducing equivalents) is transported into the mitochondrion by either the glycerol-1-phosphate transport or the aspartate-malate shuttle?

    a.   ATP   c.   NADH

    b. CO_2 d.   ADP

12. The number of cycles of the β-oxidation spiral that the (saturated) palmitic acid (C_{16}) goes through is

    a.   8   c.   7

    b.   16   d.   4

13 A main function(s) of the pentose phosphate pathway is (are)

    a.   generation of ribose   c.   generation of NADH

    b.   generation of NADPH   d.   generation of $NADP^+$

14. Carnitine acyltransferase is essential in the functioning of the

    a,   pentose phosphate pathway   c.   urea cycle

    b.   fatty acid metabolism   d.   ketone body production

15. The enzyme, pyruvate dehydrogenase, is located in:

    a.   cytoplasm   c.   mitochondrial matrix

    b.   the inner membrane of the mitochondrion   d.   cell membrane

16. The reaction in which pyruvate and CoA react to produce acetyl CoA, NADH and CO2 is:

    a.   oxidative decarboxylation   c.   oxidative phosphorylation

    b.   oxidative deamination   d.   reductive elimination

COMPLETION. Write the word, phrase or number in the blank space in answering the question.

1. The three end products of complete catabolism of proteins in the body are _____, _____, and _____.

2. Specialized cells that store fats are called _____.

3. In the catabolic metabolism of hemoglobin, the iron is taken up by _____.

4. _____ is considered a ketone body despite not having a ketone functional group.

5. An aldolase enzyme catalyzes the splitting of fructose 1, 6-bisphosphate to _____ and _____.

6. Fats are metabolized by first hydrolyzing them to _____ and _____.

7. Although _____ is a basic amino acid and occurs in the urea cycle, it is not found in proteins.

8. The β-oxidation of a C_{10} unsaturated fatty acid requires one _____ step in the process than that for the saturated C_{10} fatty acid.

9. Fatty acid oxidation occurs in the _____.

10. If the carbon skeleton of an amino acid is converted to pyruvate, it may serve in either of two roles.
    a. _____
    b. _____

11. Damaged proteins are covalently modified by reaction with another protein called _____ , after which this "flagged" or modified protein is delivered to the machinery for proteolysis, called the _____.

12. When the levels of glucose are high, fatty acid oxidation is _____. On the other hand, in starvation conditions, fatty acids are _____. However, the _____ produced from fatty acid oxidation cannot enter the TCA cycle because there is no _____occurring. As a result, _____ bodies are produced and used as a source of _____.

13. In the urea cycle, _____ is a molecule that is produced in the mitochondria and then diffuses into the (another organelle) _____. On the other hand, _____ is generated as one of the final products in the urea cycle and is produced in the _____ and reenters the _____. The urea is excreted in the _____

14. Although the activation of fatty acids occurs in the _____, the _____ reactions occur in the ____ of the _____.

15. The reaction in which pyruvate is converted to ethanol is called a _____ _____.

QUICK QUIZ
    Indicate whether the statements are true or false. The number of the question refers to the section of the book that the question was taken from.

28.2 What Are the Reactions of Glycolysis?
    (a) All steps in glycolysis release energy.
    (b) In glycolysis, one molecule of a C_6 compound is converted to two molecules of a C_3 compound.

301

(c) The conversion of 2-phosphoglycerate to phosphoenolpyruvate is a redox reaction.

(d) NAD^+ is regenerated when pyruvate is converted to lactate.

(e) Pyruvate can be converted to acetyl CoA, which then enters the citric acid cycle.

(f) Even though NADPH is needed for biosynthetic pathways, it can be harmful to red blood cells (RBC).

28.5 What Are the Reactions of β-Oxidation of Fatty Acids?

(a) The term β-oxidation refers to oxidation at the second carbon from the COOH group.

(b) FAD is the only coenzyme needed in the oxidation of fatty acids.

(c) Carnitine is a compound required for the activation of fatty acids before oxidation.

(d) Acetyl CoA is the principal product of fatty acid oxidation.

28.7 How Is the Nitrogen of Amino Acids Processed in Catabolism?

(a) The catabolism of amino acids proceeds by transamination or oxidative deamination, but not both.

(b) Glutamate plays a key role in amino acid catabolism.

(c) The urea cycle takes place in the cytoplasm.

(d) The presence of fumarate links the urea cycle and the citric acid cycle.

(e) The compound carbamoyl phosphate plays a role in nucleotide biosynthesis as well as in the urea cycle.

(f) Ornithine is an amino acid found in some proteins

28.8 How Are the Carbon Skeletons of Amino Acids Processed in Catabolism?

(a) The carbon skeletons of glucogenic amino acids do not enter the citric acid cycle.

(b) The carbon skeletons of ketogenic amino acids enter the citric acid cycle.

(c) Amino acids can be glucogenic or ketogenic, but not both.

(d) Carbon skeletons of amino acids can be used for biosynthesis.

ANSWERS TO SELF-TEST QUESTIONS

MULTIPLE CHOICE

1.	a	8.	c
2.	a	9.	c
3.	b	10.	c
4.	b	11.	c
5.	c	12.	c
6.	a	13.	a, b
7.	d	14.	b
		15.	b
		16.	a

COMPLETION

1. urea or NH_4^+, CO_2, H_2O
2. fat deposits
3. ferritin
4. β-hydroxybutyrate
5. glyceraldehyde 3-phosphate, dihydroxyacetone phosphate

6. glycerol, fatty acid
7. ornithine
8. more
9. mitochondrial matrix
10. a.   an energy source
    b.   a building block in the synthesis of glucose (and other metabolites)
11. ubiquitin, proteasome
12. inhibited, oxidized, acetyl CoA, glycolysis, ketone, energy
13. citrulline, cytoplasm, ornithine, cytoplasm, mitochondrion, urine
14. cytoplasm, β-oxidation, matrix, mitochondrion
15. reductive decarboxylation

QUICK QUIZ

28.2
    (a) False   (b) True   (c) False   (d) True
    (e) True   (f) False
28.5
    (a) True   (b) False   (c) False   (d) True
28.8
    (a) False   (b) True   (c) False   (d) True
    (e) True   (f) False
28.9
    (a) False   (b) True
    (c) False; there are three amino acids that can be both.
    (d) True

If we could not store so vital a
substance as glucose in our bodies,
we would have to be eating incessantly
to maintain a steady supply of it. That
would be a precarious existence.
- Ernest Borek
The Atoms Within Us

29

Biosynthetic Pathways

CHAPTER OBJECTIVES

After you have studied the chapter and worked the assigned exercises in the text and the study guide, you should be able to do the following.

1.   Contrast anabolic and catabolic reactions in terms of energy requirements, the form of coenzymes involved and the cellular location in which they generally take place.

2.   Explain why anabolic reaction pathways are usually not just the reverse of the corresponding catabolic pathways.

3.   Explain how the Cori cycle coordinates actions in the muscles and in the liver during strenuous exercise.

4.   Using Figure 29.1 (text), identify the major metabolic intermediates that may be used in gluconeogenesis and indicate how glycolysis differs from gluconeogenesis.

5.   Discuss the importance of UDP-glucose in carbohydrate biosynthesis.

6.   Identify the feature that is common in both fatty acid synthesis and degradation.

7.   Describe the role of acetyl CoA and malonyl CoA in fatty acid synthesis.

8.   Indicate the form of CoA and the other reactants involved in the synthesis of phosphatidate and sphingosine.

9.   Write out the reactions for the synthesis of glutamic acid from α-ketoglutaric acid and the

304

subsequent transamination reaction to form other non-essential amino acid.

10. Outline the reaction in cholesterol biosynthesis and indicate how statin drugs inhibit cholesterol synthesis.

11. Distinguish the difference between essential and nonessential amino acids.

12. Outline the photosynthetic process in terms of a light reaction and a dark reaction.

13. Define the important terms and comparisons in this chapter and give specific examples where appropriate.

IMPORTANT TERMS AND COMPARISONS

Biosynthetic (Anabolic) Pathways
Catabolic Reactions & Mitochondria
Cori Cycle (Muscle & Liver Coordination)
Gluconeogenesis
Pentose Phosphate Pathway
Photosynthesis: Photosystem I and
    Photosystem II
Light and Dark Reactions
Fatty Acid Biosynthesis
Acyl Carrier Protein (ACP)
Acyl CoA and Acetyl CoA
HMGCoA and Mevalonate
Isopentenyl Pyrophosphate
Prenylation
Essential Fatty Acid
Obesity and Body Mass Index (BMI)
Malonyl ACP
Glucogenic Amino Acid
Ribulose 1, 5-Bisphosphate Carboxylase
   -Oxygenase (RuBisCO)

Le Chatelier's Principle
Anabolic Reactions and Cytoplasm
Glycolysis
Glyocogenesis
Glycogenolysis
Chloroplasts
Chlorophyll a
Calvin Cycle
Membrane Lipid Biosynthesis
Uridine Diphosphate (UDP)
Cholesterol Biosynthesis and
    HMG Reductase
Geranyl and Farnesyl Pyrophosphates
Signal Transduction and G-proteins
Fatty Acid Synthase
Malonyl-CoA & Acetyl-CoA Carboxylase
Gluconeogenesis and Glycogenesis
Essential and Non-Essential Amino Acids
Kwashiorkor

SELF-TEST QUESTIONS

MULTIPLE CHOICE. In the following exercises, select the correct answer from the choices listed. In some cases, two or more choices will be correct.

1. Gluconeogenesis is associated with the synthesis of

    a.   glycogen   c.   glucose
    b.   glucogenic amino acids   d.   glucose-1-phosphate

2. Glycolysis and gluconeogenesis proceed in reverse directions. However, the number of points in this pathway in which there are enzymes not common to each pathway is

a. 6   c. 3
b. 8   d. 2

3. Both fatty acid synthesis and degradation have the common feature of involving

   a. ATP   c. acetyl CoA
   b. NAD⁺   d. ketogenic amino acids

4. In each cycle in fatty acid synthesis, the unit added on is a

   a. C_1 fragment   c. C_4 fragment
   b. C_2 fragment   d. C_6 fragment

5. Which of the following structures represents UDP-glucose?

a.
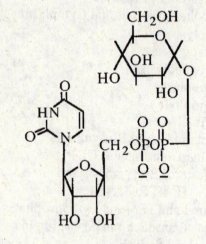

c.
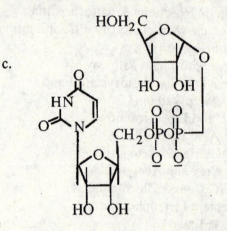

b.
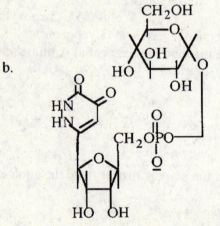

d.
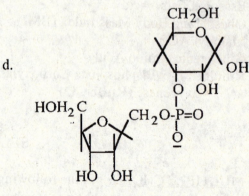

6. The amino acid produced in the following reaction is:

$$
\begin{array}{ccc}
\text{CH}_3 & & \text{CO}_2^- \\
| & & | \\
\text{C=O} & & \text{HCNH}_3^+ \quad \longrightarrow \\
| & + & | \quad\quad\quad \longleftarrow \\
\text{CO}_2^- & & \text{CH}_2 \\
& & | \\
& & \text{CH}_2 \\
& & | \\
& & \text{CO}_2^-
\end{array}
$$

a. glycine   c. arginine

b. glutamic acid   d. alanine

7. Which of the following can be used in gluconeogenesis?

    a. pyruvate   c. ketogenic amino acids

    b. lactate   d. urea

8. Which of the following are essential fatty acids?

    a. stearic acid   c. palmitic acid

    b. butyric acid   d. linoleic acid

9. The Calvin cycle is associated with

    a. the dark reaction   c. utilizes RuBisCO

    b. requires UDP-glucose   d. converts H_2O to O_2

10. The Cori cycle involves which of the following:

    a. lactate production in the muscles

    b. lactate transport from the liver to the muscles

    c. gluconeogenesis in the liver

    d. glucose transport back to the muscles for use in exercise

COMPLETION. Write the word, phrase or number in the blank space in answering the question.

1. Biosynthetic pathways require an ____ input and require _____ power in the form of NADPH.

2. Glycogen is synthesized from the activated glucose monomer, _____, which is formed in the reaction of _____ and _____.

3. The overall reaction in photosynthesis is _____.

4. Fatty acid synthesis occurs in the _____ , while fatty acid degradation takes place in the _____ .

307

5. The acetyl or malonyl group in the ACP complex is attached through a _____ atom.

6. The fatty acid component of phosphatidate is derived from the _____ which reacts with glycerol-1-phosphate.

7. The biosynthesis of cholesterol occurs in the _____ (an organ) and is assembled from _____.

8. The phosphatidylcholine in normal lungs contains predominantly _____ fatty acids.

9. Photosynthesis occurs in the _____ organelles of plants. The first reaction in photosynthesis requires energy (sunlight) and is called the _____ reaction. The four important species generated are _____ , _____ , _____ and _____ . The second reaction uses CO_2 and three of the species generated in the light reaction , _____ , _____ and _____ to produce _____ macromolecules.

10. Most anabolic reactions occur in the _____.

11. Amino acids that can be synthesized in the human body are referred to as _____ amino acids.

12. Cholesterol biosynthesis involves three intermediates that contain a C_5, C_{10} and C_{15} unit, respectively, bound directly to pyrophosphate. The names of these three units are _____ , _____ and _____.

QUICK QUIZ
Indicate whether the statements are true or false. The number of the question refers to the section of the book that the question was taken from.

29.2 How Does the Biosynthesis of Carbohydrates Take Place?
(a) Biosynthesis of carbohydrates takes place in exactly the same way in plants and animals.
(b) Gluconeogenesis is not the exact reversal of glycolysis.
(c) Some of the enzymes in gluconeogenesis differ from those in glycolysis.
(d) Adenine nucleotides are the only ones that play a role in the biosynthesis of carbohydrates.

29.4 How Does the Biosynthesis of Membrane Lipids Take Place?
(a) Glycerol 1-phosphate plays a role in the biosynthesis of membrane lipids, but sphingosine does not.
(b) Statin drugs inhibit the biosynthesis of cholesterol.
(c) All the carbon atoms in cholesterol come from the methyl group found in the acetyl group of acetyl CoA.
(d) Isoprene units play an important role in the biosynthesis of cholesterol.

29.5 How Does the Biosynthesis of Amino Acids Take Place?
(a) Some amino acids cannot be synthesized by the body and must be obtained from the diet.
(b) Transamination reactions do not play a role in the biosynthesis of amino acids.

308

ANSWERS TO SELF-TEST QUESTIONS

MULTIPLE CHOICE

1. c
2. c
3. c
4. b
5. a

6. d
7. a, b
8. d
9. a, c
10. a, c, d

COMPLETION

1. energy, reducing
2. UDP-glucose, UTP, glucose-1-phosphate

3. $n\ H_2O\ +\ n\ CO_2\ \xrightarrow{\text{sunlight}}\ (CH_2O)_n\ +\ n\ O_2$

4. cytosol, mitochondria
5. S
6. acyl CoA
7. liver, acetyl CoA
8. unsaturated
9. chloroplast,

   light, ATP, NADPH, H^+, O_2

   ATP, NADPH, H^+, carbohydrate
10. cytoplasm
11. nonessential
12. isopentenyl pyrophosphate (C_5), geranyl pyrophosphate (C_{10}), farnesyl pyrophosphate (C_{15})

QUICK QUIZ

29.2

   (a) False   (b) True   (c) True   (d) False

29.4

   (a) False   (b) True   (c) False   (d) True

29.5

   (a) True   (b) False

MyPyramid
USDA recommendations for a balanced diet.
    www.mypyramid.gov

30

Nutrition

CHAPTER OBJECTIVES

After you have studied the chapter and worked the assigned exercises in the text and the study guide, you should be able to do the following.

1. Define the term, nutrient, and classify the nutrients into the six groups.

2. Indicate what RDA and DRI stands for.

3. Explain the difference between basal caloric requirement and the normal caloric intake of active individuals.

4. Explain why exercise and a lengthy diet are usually necessary to lower weight by more than a few pounds.

5. Indicate the primary roles in the body for carbohydrates, fats and proteins.

6. Distinguish between digestible and non-digestible carbohydrates.

7. Distinguish between a complete and an incomplete protein from a nutritional standpoint.

8. Outline what occurs during the process of digestion.

9. Outline the digestion process for carbohydrates and point out the role of the most important enzymes involved.

10. Outline the digestion process for lipids and point out the role of the most important enzymes involved.

11. Outline the digestion process for proteins and point out the role of the most important enzymes involved.

12. Name the water-soluble vitamins and indicate one function for each.

13. Name the fat-soluble vitamins and indicate one function for each

14. Define the important terms and comparisons in this chapter and give specific examples where appropriate.

IMPORTANT TERMS AND COMPARISONS

Nutrients: Carbohydrates, Lipids, Proteins,
    Vitamins, Minerals and Water
Nutritional Calorie (Cal) = 1 kcal

Food Guide Pyramid
Discriminatory Curtailment Diets
Basal Calorie Requirement
 α–Amylase, β–Amylase and
   the Debranching Enzyme
Lipids and Lipases
Essential Fatty Acids: Linolenic and Linoleic Acids
Complete Protein
Protein Complementation
Kwashiorkor
Fat-Soluble Vitamins
   Vitamin A
   Vitamin D
   Vitamin E
   Vitamin K
Vitamin and Corresponding Coenzyme
Minerals
Artificial Sweeteners
Performance Enhancing Foods: Creatine
Anabolic Steroids
Olestra
Hypoglycemia and Hyperglycemia

Digestion
Dietary Reference Intakes (DRI) and
   Recommended Daily Allowance
   (RDA)
Fiber and Cellulose
Marasmus and Obesity
Body Mass Index (BMI)
Stomach Acid
Starch and Glycogen
Bile Salts
Essential Amino Acids (10)
Digestive Proteolytic Enzymes
   Chymotrypsin, Pepsin
   Trypsin, Carboxypeptidase
Water-Soluble Vitamins
   Vitamin C (Ascorbic Acid)
   Vitamin B_1 (Thiamine)
   Vitamin B_2 (Riboflavin)
   Vitamin B_6 (Pyridoxal)
   Nicotinic Acid (Niacin)
   Folic Acid
   Vitamin B_{12}
   Pantothenic Acid
   Biotin
Ergogenic Aid

SELF-TEST QUESTIONS

MULTIPLE CHOICE. In the following exercises, select the correct answer from the choices listed. More than one correct answer may be possible in these questions.

311

1. Which of the following are classified as nutrients?

    a. carbohydrates
    b. nucleic acids
    c. esters
    d. vitamins
    e. water

2. A nutrient calorie (Cal.) is equivalent to

    a. 100 calories
    b. 1 calorie
    c. 1000 calories
    d. no fixed number because it will depend on the route of digestion

3. The most prevalent form of dietary fats is (are)

    a. fatty acids
    b. cholesterol
    c. triglycerides
    d. sphingolipids

4. The essential fatty acids are

    a. linolenic and linoleic acids
    b. acetic and linoleic acids
    c. palmitic and linolenic acids
    d. stearic and linoleic acids

5. Water makes up what percent of body weight

    a. 15%
    b. 75%
    c. 35%
    d. 60%

6. A number of enzymes are necessary for the digestion of carbohydrates. The role of β-amylase is to

    a. hydrolyze the α (1-4) glycosidic bonds in an orderly fashion, cutting disaccharide maltose units one by one from the nonreducing end of the chain
    b. randomly cut the α (1-4) glycosidic bonds
    c. hydrolyze the α (1-6) glycosidic bond
    d. hydrolyze all glycosidic linkages, irrespect of the type of linkage

7. A carbohydrate that cannot be hydrolyzed in our digestive system is

    a. glycogen
    b. cellulose
    c. amylose
    d. amylopectin

8. Some of the final digestion products of lipids as a result of lipase action are

    a. glycerol
    b. choline
    c. phospholipids
    d. fatty acids

9. Digestion of proteins results from the hydrolysis of the

a. phosphodiester bond
b. peptide bond

c. glycosidic backbone
d. hemiacetal bond

10. Enzymes involved in the digestion of proteins include

a. pepsin
b. lipase

c. carboxypeptidase
d. α-amylase

11. The water-soluble vitamins include

a. vitamin B_{12}
b. biotin

c. thiamine
d. niacin

12. The only fat-soluble vitamin in the following list is

a. vitamin K
b. niacin

c. ascorbic acid
d. folic acid

COMPLETION. Write the word, phrase or number in the blank space in answering the following questions.

1. In order to lose 10 pounds of fat, a person must use up _____ Cals.

2. We must consume certain amounts of vitamins, minerals and nutrients. Typically, we must take in _____ (a quantity) of nutrients, while we need only _____ (quantity) of vitamins and minerals.

3. Most of the minerals are metal ions. Five of the metal ions are _____, _____, _____, _____, and _____.

4. Two minerals that are not metals are _____ and _____.

5. Two fat-soluble vitamins are _____ and _____.

6. Four water soluble vitamins are _____, _____, _____ and _____.

7. Linolenic and linoleic are essential fatty acids in higher animals. They both contain _____ carbons in the make-up of the fatty acid.

8. Three water soluble vitamins that are converted to coenzymes in biological reactions are _____, _____ and _____.

9. For the dietary purpose of supplying amino acids, _____ is a complete protein, while examples of proteins that are not complete are _____ and _____.

10. _____ are the most concentrated form of energy that we consume as food.

313

11. Olestra is effectively a fat since is has _____ functional groups. It is derived from the sugar, _____, in which the _____ groups have been esterified by reaction with long chain fatty acid groups.

12. Although _____ is not a nutrient, it is an important indigestible ingredient in our diets.

13. Iron usually occurs in foods in the _____ form, but must be _____ to _____ by vitamin __ to be absorbed.

14. Many athletes use the naturally occurring amino acid, _____, as a performance-enhancing food

MATCHING.   Match the term in the right-hand column with the appropriate answer in the left-column.

1.	Vitamin A	a.	Fat-soluble antioxidant
2.	Thiamine	b.	Found in leafy vegetables
3.	Ascorbic Acid	c.	Contains the metal, cobalt
4.	Vitamin K	d.	Another name for vitamin B_1
5.	Vitamin E	e.	Important in bone formation
6.	Vitamin D	f.	Prevalent in citrus fruit
7.	Folic Acid	g.	Important for blood clotting
8.	Vitamin B_{12}	h.	Important for proper vision

QUICK QUIZ

Indicate whether the statements are true or false. The number of the question refers to the section of the book that the question was taken from.

30.1 How Do We Measure Nutrition?
(a)   Nutritionists divide nutrients into 6 categories
(b)   The levels of nutrients required for a healthy diet are currently referred to as RDA's, for recommended daily allowances.
(c)   A discriminatory curtailment diet refers to a diet low in calories
(d)   Carbohydrates are at the top of the original food pyramid
(e)   The revised food pyramid distinguishes between types of carbohydrates

30.5 How Does the Body Process Dietary Protein? .
(a)   All protein digestion occurs in the stomach.
(b)   Pepsin is an enzyme that digests proteins at very low pH values.
(c)   Chymotrypsin hydrolyzes peptide bonds on the amino side of the aromatic amino acids.
(d)   There are 10 essential amino acids in humans
(e)   The protein from eggs is a complete protein
(f)   Protein complementation is a diet that mixes proteins to arrive at a complete protein.

314

ANSWERS TO SELF-TEST QUESTIONS

MULTIPLE CHOICE

1. a, d, e
2. c
3. c
4. a
5. d
6. a

7. b
8. a, b, d
9. b
10. a, c
11. a, b, c, d
12. a

COMPLETION

1. 35,000 Cal.
2. (many) grams; micro- or milligrams
3. Any five of Na, Ca, Mg, Fe, Zn, Cu, Mn, Cr, Mo or Cu
4. Any two of Cl, P, Se, I or F.
5. Any two of vitamins A, D, E or K.
6. Any four of vitamins B_1, B_2, B_6, B_{12}, niacin, folic acid, pantothenic acid, biotin, or vitamin C.
7. 18
8. Any 3 of vitamins B_1, B_2, B_6, B_{12}, niacin or folic acid.
9. casein, any two of the following; gelatin or corn, rice or wheat protein.
10. Fats
11. ester, sucrose, -OH
12. fiber
13. Fe(III), reduced, Fe(II), C
14. creatine

MATCHING

1. h
2. d
3. f
4. g
5. a
6. e
7. b
8. c

QUICK QUIZ

30.1
| (a) True | (b) False | (c) False | (d) False |
| (e) True | | | |

30.5
| (a) False | (b) True | (c) False | (d) True |
| (e) True | (f) True | | |

315

"I hate quotations. Tell me what you know."
   - Ralph Waldo Emerson

31

Immunochemistry

CHAPTER OBJECTIVES

After you have studied the chapter and worked the assigned exercises in the text and the study guide, you should be able to do the following.

1. Indicate the characteristics that distinguish innate immunity and acquired immunity.

2. Outline the role of B cells in the development of an immune response.

3. Outline the role of the T cells in the development of an immune response.

4. Indicate the characteristics of killer and memory T cells.

5. Describe the interaction between an antigen and an antibody, emphasizing the molecular interactions involved on each.

6. Describe the function of Class I and Class II major histocompatibility complexes (MHC) in the immune response.

7. Distinguish between the different classes of immunoglobulins in terms of size, function and location.

8. Characterize the variable and constant regions of an immunoglobulin, IgG, in terms of its size and structure, in addition to the function of these regions.

9. Outline three different ways that can create mutations in the V(J)D genes that help to explain antibody diversity.

10. Characterize the different types of cytokines

11. State the fundamental problem in autoimmune diseases and characterize at least one such disease.

12. Define the important terms and comparisons in this chapter and give specific examples where appropriate.

IMPORTANT TERMS AND COMPARISONS

Innate Immunity - External and Internal
Immune System: Specificity and Memory
Antibodies and Immunoglobulins
Lymphocytes - B Cells and T Cells
Antigens, Haptens and Antibodies
Cytokines - Signaling Regulatory Proteins
Macrophages and Natural Killer Cells
NO and NO Synthase
Plasma Cells
Antibodies and Immunogens
Major Histocompatibility Complex (MHC)
Classes of Immunoglobulins
Large and Small Chain (Subunits)
Antigen-Antibody Complex
Variable Region in H Chain and
    V(D)J Recombination
Affinity Maturation
Monoclonal Antibodies
Glycoprotein 120 (gp120)
Cytokines, Interleukins and Chemokines
T Cell Receptor Complex
Activating and Inhibitory Receptors
Epigenetics
Pluripotent and Multipotent
Imprinting

Acquired or Adaptive Immunity
Immunoglobulin Superfamily
Antibodies, T Cell Receptors and
    Major Histocompatibility Complexes
    (MHCs)
Lymph and Lymphoid Organs
Specific and Non-Specific Interactions
T Cell Differentiation:
    Killer, Helper and Memory T Cells
Epitope on an Antigen
MHCs - Class I and Class II
IgG, IgM, IgA, IgD and IgE
Glycoproteins
Monoclonal Antibodies
Cluster Determinant (CD) Proteins
T-Cell Receptor (TcR)
Aneuploid
Immunization
Adhesion Molecules: CD4 and CD8
HIV Virus and AIDS
Autoimmune Diseases
Hybridoma
Embryonic Stem (ES) Cells and
    Adult Stem Cells

SELF-TEST QUESTIONS

MULTIPLE CHOICE. In the following exercises, select the correct answer from the choices listed. More than one correct answer may be possible in these questions.

1. Class I MHC molecules have which of the following characteristics?. They

    a.   are single chain molecules
    b.   are transmembrane proteins

    c.   form dimers
    d.   bind to molecules in the cell that are derived from viral infection

2. An epitope is found on

317

a.	an immunoglobulin protein	c.	an antigen
b.	a T-cell surface protein	d.	a cytokine

3. The role of the MHC is to

a.	bind CD4	c.	bind to antigen in plasma
b.	stimulate cytokine formation	d.	bringing the antigen's epitope to the cell surface

4. The smallest (by MW) and most abundant antibody in the blood serum is

a.	IgG	c.	IgE
b.	IgD	d.	IgM

5. The number of polypeptide chains in an IgG is

a.	one	c.	two
b.	four	d.	six

6. The region of the polypeptide chains in antibodies that interacts directly with the antigen is

a.	N-terminal end (domain)	c.	central region only
b.	C-terminal end (domain)	d.	distributed over many regions

7. B cells are stimulated to differentiate into plasma cells by

a. exposure to cytokines.
b. antigen binding to the surface-bound antibody.
c. on interaction with macrophages.
d. after MHC display on surface.

8. Each B cell is capable of synthesizing

a. many different antibodies.
b. an unknown number of antibodies that depends on the situation.
c. only one immunoglobulin.
d. more than one, but clearly a limited number.

9. T cells bind to antigen by their surface

a.	TcR complex	c.	TcR
b.	TcR and cytokine together	d.	CD4 receptor

10. Monoclonal antibodies are derived from

a.	a single B cell	c.	a single T cell
b.	multiple B cells	d.	a specific type of macrophage

11. Cytokines are proteins that are modified by the attachment of

  a. sugar groups (glycoproteins) c. phosphate groups
  b. acetyl groups   d. methyl groups

12. The proteins that have an epitope that an antibody binds to are the

  a. macrophages   c. antigens
  b. MHCs, class II d. CD4 proteins

13. Immunoglobulins differ from T cell receptors in that:

  a. TcRs have only 2 subunits, while IgGs have 4 subunits
  b. TcRs can undergo somatic mutations, while IgG cannot
  c. IgGs bind directly with antigen, while TcRs bind only antigens that are presented by an MHC molecule
  d. IgGs are secreted from plasma cells, while TcRs are not

14. Hybridomas are sources of:

  a. TcRs   c. monoclonal antibodies
  b. gp120   d. vaccine against AIDS virus
             _____

COMPLETION. Write the word, phrase or number in the blank space in answering the following questions.

1. Myasthenia gravis is an _____ disease in which the patient develops antibodies against _____ receptors.

2. Heceptin is a _____ antibody that is used in the treatment of breast cancer and binds to the antigen called _____.

3. An immunization shot produces only a ____ ____ ____ and is not effective. If the virus infects the cells or alternatively, a second immunization shot, the so-called "booster shot", is given, this increases the response time and its effectiveness because the _____ cells divide and become both _____ and _____ cells.

4. A key difference between innate and acquired immunity is that the former exhibits no _____ or _____, while these characteristics are the hallmark of acquired immunity.

5. Of the two major types of T cells, only the _____ T cells release the protein, _____, that effectively makes a hole in the target cell.

6. The _____ cells are derived from _____ cells, which produce _____ proteins.

319

7. The HIV enters _____ T cells following the association of its specific glycoprotein, called _____, with the ____ protein on the cell.

8. HIV is an acronym for _____ _____ _____ which is an ____ virus that converts its genetic material into DNA by the use of its enzyme, _____ _____. The rapid changes or _____ that occur in the genetic material of HIV are due to the inaccuracy of _____ _____ that carries out _____.

QUICK QUIZ
Indicate whether the statements are true or false. The number of the question refers to the section of the book that the question was taken from.

31.1 How Does the Body Defend Itself from Invasion?.
(a) Dendritic cells, NK cells, and macrophages are part of the innate immunity system.
(b) Tears and mucous membranes are part of the innate immunity system.
(c) Cells of the innate immunity system are characterized by speed and memory.
(d) T cells and B cells are part of the acquired immunity system.
(e) Antibodies are molecules that elicit an immune response.
(f) Antibodies, T cell receptors and MHCs are members of the immunoglobulin superfamily.

31.2 What Organs and Cells Make Up the Immune System?
(a) NK cells and macrophages kill infected cells the same way.
(b) Antigen presenting cells attach pieces of antigens on major histocompatibility proteins.
(c) Helper, cytotoxic, and memory are all types of T cells.
(d) Lymphocytes are derived from stem cells in the bone marrow.
(e) Antibodies are produced by plasma cells once they turn into memory cells.

31.3 How Do Antigens Stimulate the Immune System?.
(a) Epitopes are molecules that elicit an immune response.
(b) The ABO blood groups are protein based antigenic determinants
(c) A foreign molecule must be of a minimum size to elicit an immune response.
(d) The smallest unit of an antigen capable of binding with an antibody is called an epitope.
(e) T cell receptors can recognize carbohydrate based antigens

31.4 What are Immunoglobulins?
(a) All immunoglobulins have the same molecular weight.
(b) IgA molecules are found mostly in secretions
(c) IgG and IgE are the most important immunoglobulins in the blood.
(d) Immunoglobulins are made up of a combination of light and heavy chains.
(e) Antibody diversity is created by the recombination of V, J, and D gene segments during B cell development.
(f) Somatic mutations that arise leading to antibody diversity are passed onto the offspring of the person that was infected.

320

ANSWERS TO SELF-TEST QUESTIONS

MULTIPLE CHOICE

1. a, b, d
2. c
3. d
4. a
5. b
6. a

7. b
8. c
9. a
10. a
11. a
12. c
13. a, c, d
14. c

COMPLETION

1. autoimmune, acetylcholine
2. monoclonal, c-erb-B2
3. short term response, memory, memory, plasma
4. specificity, memory
5. killer, perforin
6. plasma , B, antibodies
7. helper, gp120, CD4
8. Human Immunodeficiency Virus, RNA, reverse transcriptase, mutations, reverse transcriptase, replication

QUICK QUIZ

31.1

(a) True   (b) True   (c) False   (d) True
(e) False   (f) True

31.2

(a) False   (b) True   (c) True   (d) True
(e) False

31.3

(a) False   (b) False   (c) True   (d) True
(e) False

31.4

(a) False   (b) True   (c) False   (d) True
(e) True   (f) False

321